Ensuring Greater Yellowstone's Future

1997   Tim W. Clark, *Averting Extinction: Reconstructing Endangered Species Recovery*

1999   Tim W. Clark, A. Peyton Curlee, Steven C. Minta, and Peter M. Kareiva, editors: *Carnivores in Ecosystems: The Yellowstone Experience*

2000   Tim W. Clark, Andrew R. Willard, and Christina M. Cromley, editors, *Foundations of Natural Resources Policy and Management*

2002   Tim W. Clark, *The Policy Process: A Practical Guide for Natural Resource Professionals*

Note: Before 2006, Susan G. Clark published under the name Tim W. Clark.

Ensuring Greater Yellowstone's Future

Choices for

Leaders and Citizens

Susan G. Clark

Yale University Press

New Haven and London

Published with assistance from the foundation established in memory of Philip Hamilton McMillan of the Class of 1894, Yale College.

Set in Aster Roman types by Binghamton Valley Composition.

Printed in the United States of America.

Library of Congress Cataloging-in-Publication Data

Clark, Susan G., 1942–
    Ensuring greater Yellowstone's future : choices for leaders and citizens / Susan G. Clark.
        p.   cm.
    Includes bibliographical references and index.
    ISBN: 978-0-300-12422-4 (cloth : alk. paper)

    1. Yellowstone National Park—Forecasting.   2. Yellowstone National Park Region—Forecasting.   3. Yellowstone National Park—Management.   4. Landscape protection—Yellowstone National Park.   5. Ecology—Yellowstone National Park.   6. Yellowstone National Park—Environmental conditions.   7. Yellowstone National Park Region—Environmental conditions.   8. Environmental policy—Yellowstone National Park.   9. Yellowstone National Park (Agency : U.S.)   I. Title.
    F722.C565 2008
    333.7809787'5—dc22

                                                    2007043536

A catalogue record for this book is available from the British Library.

The paper in this book meets the guidelines for permanence and durability of the Committee on Production Guidelines for Book Longevity of the Council on Library Resources.

10 9 8 7 6 5 4 3 2 1

Contents

Preface

Yellowstone is one of America's most special places, and the people to whom its management is entrusted have a special responsibility. The park itself and much of the greater Yellowstone ecosystem are managed by the federal government, while individual landowners, businesses, and state and tribal governments manage the rest. Together these entities set priorities for the region, but many individuals and nongovernmental organizations also influence policies for managing the region's resources. As a result, multiple people and groups provide leadership and determine what happens to greater Yellowstone. In this book I examine the region's leadership, both in and out of government, and the policy process through which leaders seek to work together to chart a course toward sustainability. Getting "analytic traction" or a "researchable handle" on the subjects of leadership and policy process is difficult. These are complex and dynamic matters, but they are the means by which technical, cultural, and political change takes place, which in turn leads to alterations in individual behavior and institutional practices.

While focusing on leadership throughout greater Yellowstone, I will look specifically at the Greater Yellowstone Coordinating Committee (GYCC), a high-level federal committee that influences management policy and which is made up of the heads of the area's national parks, forests, and wildlife refuges. The committee's members, collectively and individually, provide leadership for the public lands in the ecosystem. They are responsible for managing these lands in perpetuity for the American people. Focusing on this committee is a way to collapse the complex narrative or discourse of leadership throughout the system into a tractable story, but this focus on one committee is not meant to be limiting. This book is intended also to offer insight into how the policy system for managing resources functions in general, and the lessons learned can be applied to all organizations operating in the region.

I analyze how the committee works in terms of its problem-solving style, patterns of cooperation, and accomplishments (to the extent that they appear in the observable record of the committee's words, actions, and documents). I am interested in how leaders organize themselves, what they talk about, and what they decide to do in response to the challenges they see. The intent is to see the issues that arise in greater Yellowstone through the eyes of decision makers who are subject to real deadlines and policy constraints.

Having spent sixteen years studying the committee and the leadership in greater Yellowstone, I offer this analysis in a constructive spirit with the greatest respect for the individuals who have served on the GYCC and the many dedicated people who have worked with them, both in and out of government. I also have deep appreciation for the difficulties of the management challenges they face in their individual parks, forests, and refuges and in the region as a whole, and for the challenges confronting leaders in greater Yellowstone's state and local governments, businesses, tribes, and NGOs.

My analysis is meant to encourage greater reflection and attention to the higher-order tasks as well as to the basic tasks required for effective leadership—not only by the region's leaders but also by all the other organizations and individuals who live there and care about the future of our natural heritage. We all need to reflect on our most basic assumptions about nature and its significance in our lives and, more practically, on the people and institutions we trust to manage greater Yellowstone sustainably. Such reflection should be "about the complicated and contradictory ways in which modern human beings conceive of their place in nature," as environmental historian William Cronon wrote in the introduction to his edited volume *Uncommon Ground*. I hope to make leaders' jobs easier by exploring the multidimensional situations in which they find themselves today. For example, they are responsible for administering federal laws yet at the same time must remain sensitive to local and regional contexts and practical politics, including the power politics of special interests. In addition, they must also respond to rapid changes underway regionally, nationally, and globally, including suburban sprawl, biodiversity loss, and climate change. Feedback can help leaders be most successful in working toward the common good.

This book's origin lies in my experiences as a wildlife researcher and conservationist in greater Yellowstone. Having lived and worked there for more than four decades, I prize the region's public lands, wildlife, and relatively pristine landscapes and ecological processes, as well as its human communities. I have deep personal and intellectual relations in this region, which to me represents one of the world's best chances to develop sustainable practices for a healthy environment and a healthy society. I have worked with government agencies at all levels and observed their struggle to manage the region with the highest standards; with the promise and foresight of the U.S. Constitution; and with the many existing federal, state, and local policy prescriptions. Theirs is a difficult job, given the complexity of the context in which they work. I have also worked with many citizens, ranchers, small businesses, and local nongovernmental organizations and have long-standing scholarly connections to the region, having taught graduate courses related to environmental issues in general and on this area in particular for many years. In my courses and seminars, I have explored the complex policy dynamics of resource management in the West, across the United States, and in other countries. Together with my students (from more than thirty-five countries), I have explored cases in diverse contexts on land, in freshwater, and in the oceans. I have also studied the ideas of other scholars, who have offered conceptual and analytic frameworks well grounded in the social and policy sciences and human experience, which can be used to understand the present Yellowstone situation. Finally, I have observed leaders in natural resource management in sixteen countries as they develop and execute policies. These experiences have introduced me to a number of insightful conceptual and practical tools and comparative examples that are pertinent to leadership and management policy in greater Yellowstone. It would be good to harvest these lessons of experience and act on them.

We are fortunate today to be in a position to debate the health of the Yellowstone region, the adequacy of existing management policy, and options for its future in a free society. It is a credit to the region's leaders over the last 130 years that we have options open to us; they preserved for us the landscape and natural heritage we cherish today. The fact is that greater Yellowstone is an enormous, intact environment of a kind found in few places on the planet. There are also knowledgeable, vibrant communities of people who care deeply about the land and its future. Yet today we face important decisions, and the results of these choices may leave future generations with fewer options than we now enjoy. It is my goal in this book to build on past accomplishments and highlight the vital elements of the long-term agendas toward which greater Yellowstone's numerous leaders are striving. Those leaders, acting on their various past agendas, have given us an outstanding natural legacy, a variety of livelihoods, and organized committed communities, but we now need to come together on a common agenda for the future.

It is my hope that this book will bring together the story of Yellowstone's history and the dynamics of its leaders and institutions in a way that permits us to see, more comprehensively than before, the patterns they create. It is essential that we try this kind of comprehensive, integrated examination periodically. It the only way to extend our meager understanding of the significant events and processes that created the idea of greater Yellowstone that we seek to perpetuate. We need to create greater understanding of our world, as well as self-knowledge, in order to strengthen our institutions and management policy in the Yellowstone region. I hope this book helps to reinvigorate our commitment and accelerate progress toward the promise that Yellowstone represents to the world.

I want to acknowledge the personal, professional, academic, and financial support I received in writing this book. There are too many names to mention them all. Denise Casey deserves special thanks for her critical advice, as does Matt Hall. I benefited from the advice of Franz Camenzind, David Cherney, Sarah Dewey, Lloyd Dorsey, Timm Kaminski, Karin McQuillan, Susan Marsh, David Mattson, Barry Reiswig, Steve Unfried, Andrew Willard, and Jason Wilmot. Larry Timchak and Mary Maj, both of whom have served as executive coordinators for the GYCC, were very generous with their time and diligent in their efforts to provide information. I would like to acknowledge the assistance of the late Laura Cuoco (1974–2005), my student and colleague at the Yale School of Forestry and Environmental Studies, who helped with data compilation. Many colleagues in the agencies, academia, and nongovernmental organizations offered discussion, critique, and encouragement. Finally, the staff and affiliates of the Northern Rockies Conservation Cooperative in Jackson, Wyoming, encouraged me through their example of working for creative, cooperative improvements in leadership and management policy in greater Yellowstone and beyond.

Ensuring Greater Yellowstone's Future

1 Leaders and Policy in a Contested Landscape

Yellowstone is the world's first and one of its greatest national parks. Its name comes from the color of the rock walls in the Grand Canyon of the Yellowstone River, a deep gorge at the heart of the park. The name is also used to refer to the mountainous, forested region surrounding the park, called the greater Yellowstone area or ecosystem. Yellowstone has also become—in addition to a place, a park, and a region—an idea about nature and our relationship to it, as well as an ethic, calling to mind our responsibility for our world. In this book I argue that the region's leaders should be actively working toward more unified, organic policies, ones that would transition the whole region toward a more sustainable management of its human and natural resources. Some people, however, would prefer to maintain the status quo or even to nudge decision making to favor business, tourism, or other special interests at the expense of sustainability. Still others want to devolve decision making to state or local levels and determine Yellowstone's fate there in a piecemeal fashion. The human dynamic in these competing visions of Yellowstone boils down to two questions: "How will we use Yellowstone?" and "Who gets to decide?" As one ecologist put it, "We now have the potential of protecting it as a functioning ecosystem, or losing it."[1] Or, as nature writer Ted Kerasote pointedly asked, "Do we want to transform the remaining small slice of the unaltered continental pie . . . into what the other 90 percent looks and sounds like?"[2]

Many policy challenges confront greater Yellowstone today, and many more are foreseeable in the not-too-distant future—the management of elk and bison, large carnivores, and transportation and recreation, to mention just a few. The responsible agencies are actively working on most of these problems in one way or another, but the challenges are numerous and growing in number, complexity, and impact. Many issues are highly contentious and not easily resolvable. Civility and social capital are being drawn down in some cases. For instance, the decision to lease 175,000 acres in the Wyoming Range of western Wyoming for oil and gas development created a firestorm of public opposition in 2004.[3] In 2005, Grand Teton National Park grappled with a transportation plan that included a highly controversial fifty miles of new bicycle paths in key wildlife habitat.[4] The U.S. Fish and Wildlife Service spent years deciding how to manage the Jackson Hole elk herd that winters on the National Elk Refuge, including controversial alternatives presented in an environmental impact statement (EIS) released in summer 2005.[5] In another case, the editor of the *Jackson Hole News and Guide* concluded that "Feds fail public" and that the supervisor of Bridger-Teton National Forest had "let down the public in a recent decision allowing construction of a restaurant in a destructive avalanche path."[6] There are hundreds of other examples. Too often government-led management efforts lead not to finding common ground but to a hardening of the thinking and actions of special interest camps. This is not good for democracy, civic engagement, or natural resources. It is certainly does not encourage public trust in government or agency leaders.

One thing is clear: people in the region are finding it difficult to identify, secure, and sustain their common interests. Interactions are often politicized and conflict-ridden as rigid ideologies crash against one another. This dynamic speaks to the inadequacy of the institutions involved, in which people and organizations too often seek decisions based on the advocacy of narrow self-interest, obfuscating democratic policy and procedures, not using their staff expertise, failing to consult with the public, or not following established policy and procedures. Although oil and gas leasing and other nuts-and-bolts issues capture attention and focus public debate, the underlying issues are nearly always about the adequacy of the governance and constitutive processes, that is, everyday decision making and foundational decision making. This mix of ordinary, governance, and constitutive challenges poses a formidable metachallenge resistant to business-as-usual kinds of solutions.

Importantly, these issues focus on the complex situation in which the region's leaders find themselves today. The job of public officials is to ensure viable, sustainable ecosystem services for human benefit. Although a large body of law and regulation is committed to achieving this goal, and a huge national investment has been made over the last century, a wide array of problems caused by humans poses major threats to this goal. A selection of these are described in this book. Added to this legacy, poorly functioning institutions fail to resolve these cumulative problems. Leadership is clearly needed, but the quality of leadership is being called into question.

Something is missing in our current deliberations and labors to secure Yellowstone's and our own future. We seem trapped in coping endlessly with innumerable, onrushing, ordinary problems and the associated advocacy of special interest politics. Our attention has been captured by the ordinary challenges in management policy, the stuff of headlines in any regional or local newspaper. The discourse about how to manage greater Yellowstone is narrowly constrained by conventional thought and action and by our structures of governance, including a glaring lack of public arenas to address these issues. There are few places where someone interested in the status of Yellowstone can go to engage like-minded people about the vital issues of governance and constitutive decision making. It seems that no one is thinking about the higher-order issues at play. There is little expression of such thinking in the public words or deeds of those who are in a position to generate or initiate discussion in the regional discourse.

The most likely candidate for attending to higher-order issues, such as problem-oriented thinking, functional problem solving, and systemswide challenges, is the Greater Yellowstone Coordinating Committee (GYCC). As the highest level, resident, federal government group affecting resource management policy for most of the greater Yellowstone area, it is the entity most likely to be successful in addressing not only ordinary problems but also governance and constitutive challenges. The GYCC is made up of individual "unit managers" who are responsible for almost all public lands (national parks, forests, and wildlife refuges) in the region and whose decisions also affect private lands and other levels of government. The members of the GYCC play a pivotal role in overall management policy. Focusing on this group can give us insight into broader leadership issues and the policy dynamic throughout greater Yellowstone. The GYCC has considerable authority and control; an increasingly unified vision of improved management policy; and a vast, largely untapped potential for filling the leadership void. In this book I look at the way the GYCC goes about understanding and addressing the challenges of managing greater Yellowstone and planning for its future, and also at new approaches that can serve leaders and citizens alike in sustaining greater Yellowstone for us all.

Both as a geographic place and as an idea, Yellowstone looms large in the thinking of people of the region, the nation, and the international community. This conclusion is inescapable when standing on top of Mount Washburn in the center of the park, taking in the sweeping vistas, and it is equally true looking at the history of the Yellowstone idea and how it has been contested since its origin. Today Yellowstone is both a debated idea and a contested landscape. People have always had different notions about what Yellowstone should be. Historian Paul Schullery observed that although the park was set up in 1872, "we have never stopped establishing Yellowstone," and we continue to discover, explore, and create it with every visit.[7] We, too, reestablish or reinvent the idea of Yellowstone continually, adapting it and making it meaningful for ourselves in our time. As each generation goes through this process, we should be asking: How are we reinventing Yellowstone at this point in time? What are the consequences of our reinvention? Is any individual or group in charge of the reinvention process? What should we be doing that we are not doing to keep us on the best reinvention track? And finally, who should decide?

In studying the GYCC, I have used the case study approach to examine the committee's behavior. The case study method is "an empirical inquiry that investigates a contemporary phenomenon within its real-life context, especially when the boundaries between phenomenon and context are not clearly evident."[8] I have attended many meetings since 1991, had many informal conversations with members, made extensive and intensive observations, read reports and minutes, traveled throughout the region many times, and discussed management policy and leadership issues with people in all sectors of society in the region and beyond. As a result, research for this book rests on grounded theory, that is, theory based on finding emergent data from a situation that is studied for a long period. In grounded theory, a theory or overall picture emerges from the situation as the patterns of evidence emerge from the data. The social sciences have been using grounded theory for decades with good result.

I base my research and analysis on the policy sciences, a widely used analytic approach that provides a highly practical methodology to study the greater Yellowstone situation.[9] The data collected from diverse sources were cross-checked to ensure reliability and validity.[10] All these methods, data, and standards were "triangulated," that is, a variety of qualitative and quantitative methods were used to collect data and empirically test findings. This approach is critical to developing a comprehensive and useful data set, especially for social science research in which it is not possible to use a controlled, positivistic experimental method. Applied rigorously, this approach has proven quite effective in the study of complex, real-world cases.

In this chapter I present an overview of leadership and resource management policy in greater Yellowstone, its current dynamics, and the context of operations. In chapter 2 I survey the challenges facing greater Yellowstone conventionally, in everyday language, and then diagnose them functionally and offer a problem-oriented overview. In chapters 3 and 4 I look at the Greater Yellowstone Coordinating Committee as an example of leadership in the region, describing the committee's problem-solving behavior, patterns of cooperation, and track record of practical accomplishments. In chapter 5 I undertake a net assessment of leadership and management policy in the region, focusing on leaders, bureaucracy, and context. Finally, in chapters 6, 7, and 8 I offer ways to improve leadership and management policy to make the transition toward sustainability. These recommendations should be of interest both to people in greater Yellowstone and to those elsewhere who are working toward sustainability in a world that is growing in complexity and uncertainty. Greater Yellowstone has a strong potential to become a laboratory for environmental policy, developing solutions that the rest of the world can use in the future.

Sharing Resources—Finding Common Ground

Greater Yellowstone is a unique geological and biological system about 300 miles from north to south and 150 miles from east to west located in northwestern Wyoming, southern Montana, and eastern Idaho (figure 1.1).[11] It is about 30,000 square miles or 19 million acres with Yellowstone National Park at its center. It is the largest geothermal basin in the world. Glaciers formed much of the present landscape. Long, cold winters and brief, cool summers characterize the region. The headwaters of three major river systems are here—Yellowstone-Missouri, Green-Colorado, and Snake-Columbia. Seven floras converge in greater Yellowstone, and the region's fauna is basically intact with ten fish species, twenty-four amphibian and reptile species, three hundred bird species, and seventy mammal species. Large herds of elk and bison and populations of grizzly bears, wolves, and mountain lions live here.

So what does it matter if scores of oil and gas wells are drilled on national forest or Bureau of Land Management (BLM) lands, if wildlife habitat is destroyed in a national park to make way for bicyclists, if elk on a national wildlife refuge are managed in a way that magnifies the risks for catastrophic disease outbreaks, or if a private restaurant is

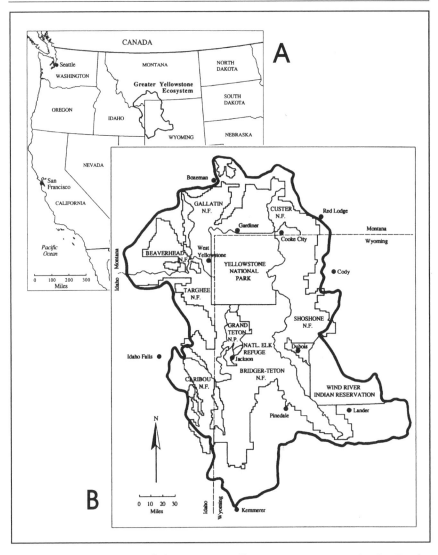

Figure 1.1. A. Location of the greater Yellowstone ecosystem in the Rocky Mountains of the United States. B. Administration of the greater Yellowstone ecosystem showing major jurisdictions.

permitted on public lands overriding public safety concerns? How should we understand what is really at stake in greater Yellowstone and what might be the consequences of the present trajectory in decision making?

One way to answer these questions is to focus on resources in functional terms. Greater Yellowstone contains shared (or "sharable") natural and cultural resources, including its land, air, water, minerals, timber, wildlife, open space, people, and institutions. These resources and the wide range of potential values they represent constitute important features of the region's human social process, and how they will be used and who will make the decisions about them are therefore subjects of great importance.[12] The kind of decision-making process and leadership that we end up with will set the patterns of use and control of these resources. Thus, the resources, the social process, the decision process, and the leaders are all critical variables in greater Yellowstone's future. Managing the region is made more complex by additional considerations, such as the scarcity of resources, changing perspectives and demands of people over time, and interdependencies and interrelationships among people. We can see these variables interact, for instance, in the claims and counterclaims made by advocacy groups about how greater Yellowstone should be managed and the adequacy of the present policies and leadership.

In the most fundamental sense, it is the job of leaders to sort out the contradictory value claims, to manage, and to make decisions in the common interest of all Americans. Managing resources is a process—not about trees, roads, or tourists, but about people struggling over time to clarify what they value and to harmonize their conflicting interests into a workable common interest. Although many people take an "objectified" view of management as manipulation of things in nature, it is in fact a human process of social interactions. It is more about people and our perspectives, values, and institutions than it is about things "out there" in nature. To understand leaders and their practices, we need to look at these larger social and decision processes of which leaders are only one part. We, the public, need to participate more effectively by better understanding both leaders' activities and the context of their operations.

People in greater Yellowstone have many interests, or demands for values, some of which overlap or conflict.[13] This milieu of public "desires" is the context in which greater Yellowstone's leaders operate, and, of course, leaders have their own interests as well. Common interests are those that are demanded by many people and benefit the entire community. The Preamble of the United States Constitution articulates on the national level a common interest "to form a more perfect Union, establish Justice, insure domestic Tranquility, provide for the common defence, promote the general Welfare, and secure the Blessings of Liberty to ourselves and our Posterity." In contrast to these interests that are clearly sought by all citizens, special interests benefit only a small number of people and disadvantage most others. This happens, for example, when officials award special concessions on public lands to a small but powerful business interest, which then abuses the land and its privilege.[14] Just such a case occurred in Teton Wilderness when outfitters placed salt baits just outside Yellowstone National Park's southern border to draw elk out of the park for their hunting clients. The Wyoming Game and Fish Department and the U.S. Forest Service encouraged this practice until other interests complained and Wilderness Watch threatened to sue the Forest Service. Such practices catering to special interests are not isolated incidents in government or in the private sector.

The Search for Sustainability

Leaders in greater Yellowstone must actively join in the effort to move toward sustainability, which in this region is often framed in the language of transboundary or ecosystem management. Clearly, much more thought and action are needed to bring this transition to reality. The U.S. National Research Council concluded in *Our Common Journey: A Transition Toward Sustainability* (1999) that "a transition is underway to a world in which human populations are more crowded, more consuming, more connected, and in many parts, more diverse, than at any time in history."[15] Concomitant with this change must be a move toward managing resources sustainably, that is, developing the "potential for a system to maintain or improve its functioning and the benefits derived from it."[16] For sustainability to be meaningful, it has to be made specific for particular places, situations, and contexts. Changes at global, national, and regional scales greatly affect what will happen in greater Yellowstone and what options leaders will have to improve management policy and to move toward sustainability.

Leaders in greater Yellowstone can draw on the experience of many others to help in their planning. Many, many books, conferences, workshops, Web pages, and demonstrations of large-scale management and policy initiatives across jurisdictional boundaries on land, in fresh water, and in the oceans attest to the growing recognition that sustainability is worth pursuing. Many efforts call for "transcending boundaries" in their search for sustainability. It is estimated that nearly seven hundred international terrestrial transboundary efforts are underway.[17] In 1997 twenty-seven of these involved three or more countries.[18] Hundreds of subnational efforts exist. More than fifty initiatives address the world's oceans and coastal areas.[19] Large-scale conservation is a major interest in many countries. New policy prescriptions are being developed and implemented in spite of overwhelming obstacles. The more practical of these new approaches may become commonplace in coming decades. Leaders in greater Yellowstone are in fact talking about some of these approaches, and they can draw on lessons from other initiatives throughout the world, many of which are well ahead of efforts here in both concept and application.

The sustainability aspiration in greater Yellowstone is intimated in forest plans and similar management unit procedures and often in terms of transboundary management, rather than sustainability per se. "Transboundary" is used here to include management across boundaries within a country, as in greater Yellowstone, as well as among two or more countries.[20] There are many kinds of boundaries besides geographic or jurisdictional ones that must be overcome, including agency, disciplinary, cognitive, organizational, professional, and parochial boundaries. The job of overcoming these boundaries is challenging in the extreme. It has been pointed out that improving management means creating "public value" or, in other words, crafting common interest outcomes over and above what we enjoy now.[21] Transboundary management promises this added public value. Such efforts seek to serve common interests by integrating historically separate social, political, and proprietary entities. Integration has been a growing endeavor throughout the world since World War II, and it currently figures prominently in discussions about natural resource management.[22] When adjacent organizations manage a shared resource, it behooves them to cooperate to the maximum extent possible. This requires initiating and reshaping the way they collectively make decisions.[23] The risk inherent in coordination, however, includes overemphasis on uniformity and a corresponding loss of creativity and flexibility. Leaders in greater Yellowstone must be vigilant and balance these concerns.

In a functional sense, transboundary management is about transforming our decision-making activities. Leaders must be fully aware of what these challenges entail, and they must be knowledgeable and skilled in order to meet them. Often this is not the case because leaders are selected, rewarded, and promoted for maintaining the status quo, not for improving on it. Most governing bodies in transboundary management programs are made up of relatively autonomous nations, regions, or organizations, each of which values its own authority and control. This is true in greater Yellowstone where local, state, federal, and tribal governments, as well as numerous people—individually and collectively—manage the region, sometimes at cross purposes. If we are to achieve genuine transboundary management, new decision-making mechanisms and rules will be required. Yet very few leaders appear to understand this necessity, talk about it openly, or do much about it.

Conventional approaches and failure to deal with larger issues will certainly undercut otherwise innovative transboundary management initiatives. The needs and problems confronting greater Yellowstone will not be resolved by continued use of everyday, routine notions, language, and methods and a continued focus on ordinary challenges on a case-by-case basis. Leaders often assume that problems can be fixed if they just scale up their tried-and-true management approaches to larger geographic areas and "coordinate." Technical fixes are then tried, more money and authority are sought, and efforts to "educate" or "involve" the public are folded in, often in incomplete or token ways. Conventional approaches also assume that existing top-down bureaucratic methods and operating protocols, such as "scientific management," using conformist norms and "expert" professionals, are adequate to engage the issues, even at large spatial scales. In fact, they are not. Consequently, many of the operating assumptions that come with this package of conventional responses (i.e., "scientific management") remain problematic and go unexamined. This is no surprise since those who most often lead transboundary efforts, namely, heads of government agencies and traditional professionals, have evolved institutionally, professionally, and cognitively to see the challenges and responses in certain traditional, standardized ways. Such thinking "inside the box" simply extends existing boundaries of thought, discourse, action, and institutional arrangements to larger scales. It does not fundamentally address the larger decision-making problems.[24] Of course, in some cases, these default or status quo approaches do achieve results, depending on the kind of ordinary problem at hand, but too often they lead to weak or failed programs or, predictably, they create new problems.

It is natural for people to begin at whatever level of understanding, insight, and resources they have at the time. The organizations people work for and the institutions they are part of set sideboards or limits for thought and action. Leaders as well as workers are subject to these boundaries. After all, they inherited them, probably helped build and refine them, and developed loyalty to them. In fact, one of the reasons leaders rise to the top is because they are skilled at organizational and institutional maintenance. It is to be expected that they would apply existing arrangements to new challenges as they arise, without questioning or deviating from established formulas. What is needed, however, are efforts to move leaders "outside the box" toward genuinely integrated transboundary thinking, learning, and practice. The challenges must be clearly defined, and more comprehensive, effective programs put in place.[25] Comparisons with other approaches and efforts elsewhere could also provide lessons to improve transboundary management policy.[26]

The GYCC claims it is "transcending boundaries" in its efforts to manage greater Yellowstone better. But we need to ask which boundaries are being transcended or should be transcended. There are natural boundaries of vegetation communities and wildlife ranges and geological boundaries of mountain ranges, creating watersheds and ecosystems. There are also jurisdictional boundaries subdividing greater Yellowstone into parks, forests, refuges, state lands, private property, Indian reservations, and other categories, each with its own legal mandates and management directives, each with its own authority and control, largely autonomous from all the rest. There are many other boundaries too—political, professional, and cognitive. Few officials within the federal agencies are willing to speak up about managing ordinary problems, let alone governance issues, beyond the borders of their own units. Who, then, is responsible for the even larger constitutive decision process that crosses all boundaries? Just whose job is it to keep track of the hundreds of ordinary cases in Yellowstone as well as the effectiveness of the overall governance processes? Who is accountable for determining how well we are doing in constitutive terms? And if the constitutive process is found lacking, who is supposed to lead us to a more adaptive, sustainable constitutive process?

Moreover, the literature presents conflicting conceptions of the goals of transboundary (and ecosystem) management. For example, T. R. Stanley, an ecological scientist, identifies two prevalent views (anthropologists call these "myths").[27] One view is *biocentric* in that "human use is considered a goal, perhaps achievable or perhaps not, which is constrained by the overall goal of protecting ecological integrity." The other is *anthropocentric,* presuming that "we can continue to manipulate and manage ecosystems to satisfy human needs and desires while protecting ecosystem integrity." In the anthropocentric view, Stanley says, "protecting ecosystem integrity does not take priority over human use." These two myths are in conflict. In an often-cited example of the biocentric view, Edward Grumbine, a conservation biologist and environmental educator, identified five frequently endorsed goals of ecosystem management: (1) maintain viable populations of all native species in situ; (2) represent, within protected areas, all native ecosystem types across their natural range of variation; (3) maintain evolutionary and ecological processes (that is, disturbance regimes, hydrological processes, nutrient cycles, etc.); (4) manage over periods of time long enough to maintain the evolutionary potential of species and ecosystems;

and (5) accommodate human use and occupancy within these con-straints.[28] But how should practitioners decide to accommodate human uses and occupancy, given ecological limits?[29] Or should they put hu-man requirements and desires first, at least as a procedural matter, as recommended by advocates of anthropocentric management?[30]

It is not possible to draw up specific goals for transboundary man-agement in the abstract. Because goals that are specific and appropriate for one context may be inappropriate in another context, this process must, instead, take place in each individual situation.[31] For instance, it is simply not possible to maintain all evolutionary and ecological processes in a setting where base metal mining is taking place. Yet modern society requires metals. Conversely, it is not possible to mine metals in a setting where all evolutionary and ecological processes are to be maintained, at least in the short to medium term. Yet there is also a demand to maintain evolutionary and ecological processes.

In summary, nowhere is the need for better transboundary man-agement policy more evident than in the greater Yellowstone arena. Meeting this need, however, will require a new kind of leader not only in government but also in the scores of environmental nongovernmen-tal organizations (NGOs) that are active in the region, and in the many businesses and other interests that want to conserve resources and improve their management. In some cases these various entities have formed powerful alliances (for example, the Yellowstone Business Council or the Yellowstone to Yukon Conservation Initiative) that pro-vide new venues for leadership. Finally, there are also many individuals who are pushing for a transition toward sustainability though trans-boundary management. This collective effort, though often underorga-nized, is well under way. It remains to be seen what its net effect will be.

The Greater Yellowstone Coordinating Committee

The Greater Yellowstone Coordinating Committee is made up of nine unit managers and two regional managers who are directly in charge of 14 million of greater Yellowstone's total of 19 million acres (figure 1.1). The committee's members manage two national parks (Yellow-stone and Grand Teton), six national forests (Bridger-Teton, Shoshone, Gallatin, Beaverhead, Custer, and Caribou-Targhee), and three national wildlife refuges (National Elk, Grays Lake, and Red Rock Lakes). To-gether they spend more than 100 million dollars annually and employ about 1,500 full-time people. They have been given the authority and control over most of the region's federal lands and are charged with managing them in the public interest.

One of the GYCC's goals is "providing leadership in making coordinated decisions that serve the public and help sustain the resources," according to official documents. In 1994 it set the goal of insuring "coordination of strategies and practices across all national parks and forest units in greater Yellowstone." In 1995 the committee said it would use an "ecosystem perspective . . . in all activities." In 1997 it listed thirty-three specific "issues" or problems that needed attention, including protecting elk, large carnivores, recycling, recreational development, air and water quality, noise and overflights, highways, livestock grazing, geothermal issues, and wildlife diseases. That same year the committee also said it would "provide a forum for the interaction [of the agencies at all levels] and private organizations and the public." By 2001 it had shifted and expanded its focus to problems with land use patterns, soil, noxious weeds, trout, imperiled species, waterways, recreation, roadless areas, fire management, data sharing, grizzly bears, and diseases, among other issues. Overall, the committee said it would discuss, encourage, inform, educate, research, plan, find consensus, use a land ethic, pool resources, coordinate, monitor progress, do exchanges, evaluate, document what has happened, and implement ways and means to meet goals and overcome problems.

In framing and addressing these ongoing challenges, agencies in greater Yellowstone began earnest work toward developing an ecosystem management perspective in the mid-1990s. The committee adopted a new philosophic paradigm, at least symbolically, in accordance with new directives during the Clinton administration to the Park Service, Forest Service, and U.S. Fish and Wildlife Service. Ecosystem management represents a genuinely new policy concept and precept for coordination among the agencies. Randall Caldwell, a natural resource policy scholar, described the significance of this new way of thinking: "Application of the ecosystem concept implies a whole new way of organizing man's relations with the natural world. An ecosystem approach to public policy implies fundamental changes in the rights and responsibilities of individuals and corporations in the possession and use of land."[32] The strategic goal of such an approach, according to H. Michael Rauscher, a specialist in decision-support systems, is in general to "find a sensible middle ground between ensuring long-term protection of the environment while allowing an increasing population to use its natural resources for maintaining and improving human life."[33] Rauscher went on to say that trying to apply ecosystem management is a "wicked" problem that embodies "ambiguities, conflicts, internal inconsistencies, unknown but large costs, lack of organized approaches, institutional shock and confusion, lack of scientific understanding of management consequences, and turbulent, rapidly changing power centers."[34]

The GYCC has chosen to use an ecosystem management approach to address the challenges as it sees them. It wants to transcend "boundaries" that are now seen as problematic for agency and interagency operations. The concept of ecosystem management has evolved from an interesting application of ecological theory to become a dominant model for natural resource management in the United States, at least rhetorically, according to Murray Rutherford, a policy scientist who studied ecosystem management in the Yellowstone region.[35] All the major federal agencies (National Park Service, U.S. Forest Service, U.S. Bureau of Land Management, and U.S. Fish and Wildlife Service) have adopted it in some form, as have many other federal, state, and local agencies, nongovernmental organizations, and private property owners.[36] In a 1994 survey, Steven Yaffee, a professor in natural resource management, and his students identified 619 sites in the United States where ecosystem management was being carried out, although it was often under other names, such as "integrated management," "watershed management," or "community-based management."[37] This evolution has taken place without significant changes in legislation or official agency mandates, although new requirements added since the late 1970s to the laws controlling administration of public properties and mandating various kinds of management plans may have accelerated the trend.[38] Clearly, ecosystem management is a powerful concept that has influenced many people. For nearly two decades, there has been discussion about its utility, but whether the GYCC's understanding and application of ecosystem management can overcome the challenges and boundary problems remains to be seen.

Even though the GYCC was instituted as a committee that was not intended to make decisions, from a functional standpoint this has not been the case. The group and its members are, in fact, an integral part of the continuous flow of social and decision processes in the region. Although in theory the GYCC is not responsible for the governance or constitutive structure within which it operates, it is in fact more aware of the issues affecting that structure than are the administrative or legislative entities to which it reports. The committee must address all the ordinary management challenges and at the same time find ways to assist in resolving interdependent governance and constitutive challenges. It has to take care of problems using routine standard operating procedures, but it must also take on the higher-order integrative and strategic functions of high-level leadership. This requires that it be creative and flexible.[39]

The challenges before the GYCC are huge, but so are the options. The agencies at the national level, the executive departments of which they are a part, and the congressional oversight committees have, individually or collectively, all seemed unable to accomplish the higher-order leadership tasks needed to ensure a sustainable future for greater Yellowstone. The GYCC is uniquely positioned among governmental groups to assume the missing leadership role. The GYCC has the authority and control to act in ways far above and beyond any narrow bureaucratic definition of its role. It has the opportunity to achieve exceptional leadership and accomplishment by taking on a special role for itself beyond what is was originally constituted to do. The committee is made up of individuals with unique perspectives and enormous cumulative experience. It is hard to imagine, in fact, a single group with greater potential and position to make a big difference. What is needed is for the GYCC to realize this and seize the unique opportunity before it.

Leadership—Promise and Options

Some people are leaders because of the job titles they hold in organizations. Others may be highly skilled in leadership but do not hold high-level positions. In some cases, position and skill do match up. In any event, it is important to identify the skills essential to true leadership, which include systematic or critical thinking (the intellectual task of orienting to problems realistically and efficiently so as to find the most effectual solutions), observation (finding problems or "targets" to focus on using extensive and intensive methods), management (interacting with people, individually and in groups, through public education, diplomacy, or writing that causes change), and mastery of technical subjects (for example, geographic information systems or technical fields, such as social or ecological methods).[40] For the GYCC to be successful, it must function well as a critical thinking body. Its leaders must observe accurately and have in-depth knowledge of the region and its history, organization, and political dynamics. They must also manage the ongoing social and political dynamics in the common interest and set up effective decision processes so that important decisions can be made in a smart and timely fashion. Finally, they must also be knowledgeable about many technical natural resource issues.

In addition to possessing identifiable leadership skills, GYCC leaders will be most successful if their style of leadership is productive. Among relevant dimensions of leadership styles are democratic versus authoritarian, task-oriented versus human relations-oriented, charismatic versus noncharismatic, and transformational versus transactional.[41] Although these categorizations do not capture all aspects of leadership styles, they do include many of the formal and informal factors actually at play. For example, James Burns, a historian and political scientist, describes the *transactional* and *transformative* styles.[42] Transactional leadership addresses the day-to-day challenges, in a piecemeal way, seeking efficiencies at the margins, a coping strategy that ignores myriad interrelated challenges and merely carries out business case-by-case.[43] Transactional leaders tend to avoid the vital higher-order thinking and policy reflection required for effective leadership in our complex, rapidly changing times. Transformative leadership, in contrast, brings about new, higher levels of operation, not only by bringing about new practices but also by inviting staff and citizens to engage morally in efforts to improve performance. To become this kind of leader requires moving beyond the transactional leadership that characterizes most administrators, agencies, and organizations today.

Effective leadership also calls for cooperation, critical in tasks such as ecosystem and transboundary management. A great many people must work in concert in greater Yellowstone's 19 million acres in order to find a coherent management policy that moves the region toward sustainability. Leaders must also carry out on-the-ground applications to demonstrate success so that everyone can see just what they are trying to achieve.

What happens to Yellowstone—as a park, a region, an ecosystem, and an idea—now and in the future is the responsibility of the region's leaders and all of us who live here and care about the land, its resources, and its people. Our leaders must be insightful, well grounded, and right in their critical judgments. Finally, they must practice proactive learning because there is no blueprint to follow along the path of transformation.

The Context of Greater Yellowstone

Greater Yellowstone as it exists today is interconnected with and thus affected by a range of contexts, from those of a global nature, to the national level, to concerns of a regional nature. Understanding past trends, current dynamics, and likely future developments at these various levels is essential to appreciate the challenges in greater Yellowstone. Leaders must be able to function effectively within these contexts.

Global

The world is experiencing dramatic change, both beneficial and harmful.[44] These changes affect how the United States will use natural resources, including those in the Yellowstone region. Dramatic technological changes are causing equally profound social transformations almost everywhere. On the social front, "no century in recorded history has experienced so many social transformations and such radical ones as the twentieth century," notes Peter Drucker, a professor and consultant.[45] The new century is shaping up to bring even greater change. Lester Thurow, professor of management and economics at the Massachusetts Institute of Technology, says that the old foundations of success in our culture are now gone. In the past, for all of human history, success rested on controlling natural resources; now it rests on "knowledge."[46]

Human impacts on global environments are also tremendous. International law professors J. Allan Beesley and Myres McDougal have noted that "highly destructive, and sometimes irreversible, damage is being done to all the resources of our global environment at an accelerating rate."[47] Some vital resources are being exhausted. Open spaces are shrinking. Agricultural lands are becoming deserts. Congestion and deterioration of urban areas is on the rise. Species extinctions are increasing rapidly. Natural beauty is being destroyed. In the last twenty-five years, global population increased 50 percent from four to six billion.[48] The world economy grew to six trillion dollars by 1950; it now grows by a like amount every five to ten years.[49] Across the globe from 1980 to 2000 economic output was up 75 percent, energy use up 40 percent, meat consumption up 70 percent, auto production up 45 percent, and paper use up 90 percent.[50] In the last century many previously unknown, human-caused, environmental problems surfaced, such as those related to the chemical and nuclear industries. Billions of pounds of pesticides are used annually. Our use of fossil fuels, combined with deforestation, has increased atmospheric carbon dioxide in ways that trap the sun's heat by 32 percent more than would occur otherwise. In 2001 fifteen hundred of the world's leading scientists, including a majority of Nobel Prize winners, declared, "The earth is finite. Its ability to absorb wastes and destructive effluents is finite. Its ability to provide food and energy is finite. Its ability to provide for the growing number of people is finite. And we are fast approaching many of the earth's limits. Current economic practices which damage the environment, in both developed and underdeveloped nations, cannot be continued without the risk that vital global systems will be damaged beyond repair."[51]

There is a global "environmental debt" building, the implications of which far exceed those of national debts and reversal of which will be infinitely more difficult. As Herbert Bormann, a member of the U.S. National Academy of Sciences and professor of ecosystem ecology, noted, "This debt also borrows from humanity's future."[52]

Ecosystems and biological diversity (the variety of plants and animals in a region and their ecological interdependence) are being degraded or destroyed at unprecedented scales and rates.[53] We depend for our very lives on ecosystems, "the productive engines of the planet—communities of species that interact with each other and with the physical setting they live in. They surround us as forests, grasslands, rivers, coastal and deep-sea waters, islands, mountains—even cities."[54] The earth's rich variety of plant and animal species and most of its ecosystems are dangerously threatened. Threats to species and ecosystems come from land use conversion, land degradation, fresh water shortages, watercourse modifications, invasive species, overharvesting, climate change, ozone depletion, and pollution. Globalization, the increasing economic integration and interdependence of countries, is both a cause of problems and also a possible solution to many of these same problems.

What happens in the rest of the world—socially, politically, economically, environmentally—affects what happens in greater Yellowstone. Yellowstone is still largely intact, and many people are concerned about its future. Places like greater Yellowstone present our best opportunities to reverse harmful trends, find a path toward sustainability, and create an exemplar of how the world can save its wild places.

National

Dramatic change that directly and indirectly affects greater Yellowstone is also evident at the national level. In 1975 Harold Lasswell, a social scientist, offered one of the most insightful accounts of America and its sociopolitical future in *The Future of Government and Politics in the United States.*[55] At that time, when the cold war was still at the center of our thinking, he forecast the rise of terrorism. He called for significantly upgraded education as the principal means to instill "democratic character" in citizens and to develop the knowledge and skills of responsible citizenship. Niall Ferguson, Oxford University historian, has forecast a grim future for us all with more "terrorism, more economic decay, . . . and more cultural fragmentation."[56] Our increasing preoccupation with terrorism will distract us from constructive solutions to pressing problems, including environmental ones. Matthew Brzezinski, newspaper analyst, suggests a future scenario for America as an "Israelized" and militarized nation with layers of costly security filling all public spaces.[57] This future is not unlikely. People's expectations (what they think is going to happen), identities (who or what they are loyal to), and demands (what they want) are shifting in response to sociopolitical change. Everyone today has an increased expectation of violence and insecurity. The rapid change we can expect in the future will produce a steady stream of disaffected citizens, many of whom have the potential to be dangerous. The job ahead for us all is to ease anxieties and provide assurance that the future can be made secure. Security means not only a psychological feeling of safety for individuals, but also evidence that society's institutions are adaptable and functioning well. A high-quality environment and a sustainable flow of resources are essential ingredients in our national and environmental security.

How these changes will affect the system of public order and the natural resources in the United States, indeed our very concept of citizenship, remains to be seen. Michael Sandel, professor of government at Harvard, argues that both liberals and conservatives, despite disagreements, share an impoverished view of what citizenship is, which leaves them unable to address the anxiety and frustrations that are widespread in our land.[58] He says that if American politics is to recover its health, that is, its "civic voice," Americans must find a way to debate questions of deliberative democracy. Researchers Benjamin Schwarz and Christopher Layne argue that American foreign policy since World War II has sought to prevent the emergence of other great powers, a strategy that has proven burdensome, futile, and increasingly risky.[59] The United States, they say, will be more secure and the world more stable if we change our policy to let other countries take care of themselves.

Turning to the national context for environmental concerns in America today, the rate at which we are destroying various environments is alarming. We have transformed 99 percent of the original tall grass prairie, 95 percent of the original primary forest in the contiguous states, 90 percent of old-growth forest of the Pacific Northwest, drained or filled 50 percent of original wetland areas, and likely rendered five hundred species extinct.[60] This national dynamic also affects what happens in greater Yellowstone.

Regional

Greater Yellowstone cannot stand apart from many of the rapid changes taking place beyond its borders. A great influx of new residents, tourists, and associated developments are putting diverse demands on the region's public lands.[61] Since Euro-Americans first explored the region two centuries ago, human occupancy and use have increased dramatically. There was little white settlement before the establishment of Yellowstone National Park in 1872. The park was among the first jurisdictions imposed on the area less than a decade after the creation of the Wyoming, Montana, and Idaho territories; over the next few decades other territorial jurisdictions (for example, states and departments) were imposed, and private interests began the process of organizing and using the region's natural resources. The years from 1917 until the 1980s saw heavy resource extraction. Although well established by World War I, ranching, mining, logging, and related activities spread rapidly and intensively from the region's periphery toward its core. In the last three decades of the twentieth century environmental regulation took hold, and a more integrated ecosystem management approach emerged, involving many scientific, policy, and organizational changes. Nongovernmental organizations have been instrumental in bringing about these changes. Today, the region's unique assemblage of geological, geothermal, and biotic features attracts about 10 million visitors annually, while about 250,000 permanent residents live in the region. Both visitors and residents increasingly demand a better quality of life, a trend that reflects changing global and American priorities, including specifically a desire to escape from the mounting stresses of modern life.[62] The last decade has seen more lawsuits than ever before concerning public land management as these priorities collide with the previously dominant extractive interests. As a result, citizens, NGOs, and government entities are seeking "conflict resolution" and "public participation" to address many problems.

Increasing numbers of people in the region and expanding consumption patterns are threatening the unique qualities—scenery, wildlife, and geothermal features—of greater Yellowstone. Collectively, the ways we live our lives and consume goods very much determine what happens to the land and its resources. Ironically, the very institutions and people who manage and enjoy the region are also part of the problem. Fixed bureaucratic interorganizational relationships, overreliance on traditional scientific and disciplinary problem-solving frameworks, and unexamined standard operating procedures have created a rigidity that prevents institutional adaptation. Historically, the region has tended to be politically conservative, and elected officials often have poor environmental voting records.[63] In recent years, however, some officials, managers, and citizens have sought to improve intergovernmental coordination, democratic responsiveness, and adaptability.[64] Ultimately, the region's institutions and people must also be the source of innovations for improvement.

Change in greater Yellowstone will likely accelerate and the context will become more complex. There is likely to be growing pluralism, more diverse value demands, more organized interest groups, increased demands for market solutions to problems, more calls for private-public partnerships, and growing tension between state and federal governments. At the same time, there will be more pleas for effective conflict resolution and increased citizen participation in public policy processes. At present, there is no comprehensive contextual map that identifies key trends in the region, explains the reasons for those trends, and projects them into the future. This lack of a common, shared contextual map perpetuates unproductive dialogue, conflict, and fragmentation in perspectives and institutional divisions.

Three Classes of Problems

Problems can be effectively addressed only if leaders and citizens know what they are. This seems obvious, but too often we misconstrue a problem, identify the wrong problem, see only part of a problem, or overlook it entirely.[65] Accurately defining problems becomes easier if we can classify or organize them. The hundreds of specific problems in greater Yellowstone can be classified into three basic but interconnected kinds of problems.

Categorization captures and organizes a wide array of problems into smaller, more manageable sets. It allows us to start communicating about the challenges in a more organized and comprehensive way and can help us solve them. It helps to clarify, and might help to narrow, the widening gap between what is known about policy making and how management is actually done in greater Yellowstone. Better conceptual clarity and functional, appropriate language will help leaders and citizens to talk and do something about the challenges they face. Correctly diagnosing problems is the first step in addressing them successfully. If society and its leaders cannot achieve accurate diagnosis or realistic problem definition, then they find themselves in a genuine dilemma, whereby neither problems nor their solutions are understandable.[66] What follows is a description of a widely accepted and very useful methodology of dividing problems into three categories: ordinary, governance, and constitutive.

Ordinary Problems

Some of the problems confronting the Yellowstone region concern tangible, ordinary things that the region's managers deal with daily.[67] These problems are important in and of themselves, but they are usually handled on a case-by-case basis, which diverts leaders from looking at the adequacy of the decision-making process. Examples of ordinary problems are what to do about domestic cattle killed by wolves, the decline of whitebark pine, and the increasing number of visitors to the parks and forests. They are relatively easy to grasp for most people and are the main level of focus of newspaper coverage, the annual reports of Yellowstone National Park and other agencies, and NGO lists of programmatic issues. No organization provides an overview or analysis of the many ordinary problems or their interrelationships. But focusing on a list of seemingly unconnected ordinary issues isn't the best way to understand the scope of the challenges, nor does it help us develop comprehensive strategies to resolve them.

There are hundreds of ordinary natural resource problems being worked on each day in greater Yellowstone. Ordinary challenges are objectified—the problem is thought of as being out there in "nature" and usually caused either by nature being "out of balance" or by "misguided" people. The problems are taken as objects and then made into targets of management manipulation (tearing down fences that block migration routes, putting up signs to keep people off big game winter range, installing electric fences to keep bears away from beehives, or "forcing" or "educating" people to behave). This approach to understanding problems results from a technical, scientific management viewpoint that emphasizes technical fixes—more surveys, more science, or more regulation—as solutions. Federal and state agencies deal with ordinary challenges technically through their professional staffs and standard operating procedures. Americans tend to trust that all problems can be solved, and most people share this view of the objectivity of ordinary problems and the plausibility of technical solutions. Their expectations are that the agencies and technical professionals can and will deal with these problems effectively.

Governance Problems

The second class of challenges has to do with decision-making processes or *governance*. Controversies that we see played out on the ground, in the media, in the courts, or in science advocacy—using the language of science, management, or politics—really at heart concern the functioning of the governance process. Often the "problem" in managing resources lies not with the outside, objective element, such as wildlife or rivers, but with this decision-making process itself. Resolving ordinary problems requires a long stream of decision-making activities. These include realizing that a certain situation constitutes a problem for some people, trying to figure out what happened to create the problem, debating the possible causes and conditions that created it, proposing and promoting possible solutions, evaluating the possible extraneous effects of the proposed solutions, choosing one course of action and setting that solution in motion, enforcing new regulations and having them interpreted by the courts, and, finally, evaluating how well the solution actually solved the problem. Sometimes people get stuck in one of these activities, such as calling for ever more research or endlessly debating potential solutions, or they select one solution without setting up the organizational apparatus to make it work, or they fail to fund it adequately, or they refuse to evaluate its effectiveness. Each decision-making activity has a whole set of concomitant pitfalls, potential dangers, or sources of error that, given time, create very real problems for resource management. Governance problems are less visible to most people, although they often have a general sense that something is wrong, and they usually have difficulty characterizing the problem or thinking about how to correct it.

Almost every ordinary problem or situation we deal with today exists because of past decisions. For instance, the decision decades ago to suppress all forest fires led to a number of problems in forest management (for example, fire frequency and size, disease management), which are now being addressed through new policy prescriptions. These, in turn, may lead to still other unforeseen problems that will have to be dealt with in the future. Deciding how to use greater Yellowstone and who gets to make the decisions is necessarily a value-laden process, and because these are matters of values and not technical facts, we will always be confronted with issues of politics and the negotiation of value positions. But for managers who routinely rely on technical fixes to solve problems, the presence of values in decision making amounts to *unjustified politics*. In short, some managers try to limit their own scope of action by restricting their understanding to fit the narrowly bounded worldview and discourse of their agencies and their professions. They circumscribe their existing identity and professional practice to fit their "objectified" view of problems, nature, and social and decision processes. They promote the status quo. To them, considering values is an unwarranted intrusion into their discretion and professional norms. The challenge for such managers is to think creatively about values and to find common ground within communities with diverse value outlooks. As Hanna Cortner and Margaret Moote note in their book *The Politics of Ecosystem Management,* "whether society can move toward ecological sustainability will depend on the health of our governance process."[68] A growing number of people are focusing on improving the governance process to find common ground in managing natural resources, including in the greater Yellowstone area.[69] The GYCC is itself an outgrowth of this new focus on governance and a laboratory in which to test its effectiveness.

Governance problems are not easy for the agencies (or anyone else) to deal with directly. Functionally, each part of our governance machinery has a distinct mandate and jurisdiction. Every level of government, as well as private or nongovernmental groups that educate, litigate, lobby for special interests, or watchdog industry or government, is all part of the governance process. Although a growing number of people recognize that many of greater Yellowstone's problems are really about the governance process, they are generally less clear about what to do to fix such problems. Yet, some are demanding new ways to make decisions, that is, new governance mechanisms, such as community-based problem solving. We can certainly improve resource management by upgrading our decision-making processes and skills. But decision making seldom produces a complete or permanent solution to problems of governance. The governance process follows rules that are set by society, and thus deeper problems of governance can be adjusted only by changing those societal rules through what are commonly described as constitutive processes.

Constitutive Problems

The third class of problem in greater Yellowstone is even more foundational and important than either ordinary or governance problems. It is constitutive, that is, it has to do with the persistent patterns of conduct—and the rules underlying those patterns—that society uses to address all its ordinary and governance challenges.[70] The constitutive process specifies, changes, or adjusts the rules about how decisions should be made. These rules-for-making-rules are not always clear, explicit, or even stated. They lie buried deep in our culture and in the system of public order that we all live under and depend on. Sometimes we have to infer these rules from how our institutions are organized and operated and from the outcomes of governance processes. Problems in constitutive decision making can best be seen through how we manage ordinary things and how decision making or governance is conducted over time. For example, the long-term patterns of decision making in oil and gas leasing, transportation, wildlife management, and special use permits on public lands create a clear picture of the basic kinds of processes and problems.

Constitutive decisions are more about the adequacy of the overall "system" composed of our institutions and our grounding philosophy. They are about decision making itself: who may make which decisions, when, and how; who gets to perform the various functions of decision making; the general principles or policies about decision making; the places and formalities within which decisions are made; symbolic as well as material resources that may be used in decision making; and the ways in which human and natural resources are useful or not. More technically, the constitutive process of a community can be characterized as the decisions that identify authoritative decision makers, specify and clarify basic community policies, establish appropriate structures of authority, determine who has power for sanctioning purposes, authorize procedures for making the different kinds of decisions, and secure the continuous performance of all the different kinds of decision functions necessary to making and administering general community policy.[71] For example, once Congress passed the Endangered Species Act (a constitutive act), the agencies had to come up with rules and regulations about how the new act would be implemented. Once rules and regulations were in place, they could be carried out through more or less routine decision making. It is not easy to see the constitutive decision process unfold or to pinpoint a specific decision process or any conscious or deliberate effort on the part of people to be constitutive, yet it is real and vitally important.

Many people do not have a concept of the constitutive process and thus do not understand its importance, but constitutive problems are just as real as ordinary management problems or governance challenges, and they have profound, fundamental, and wide-ranging consequences. A constitutive understanding draws our attention to social and decision processes and away from objectified nature and technical fixes. Without such understanding, it is impossible to clarify a comprehensive and workable means to influence the constitutive process. Unfortunately, many parochializing factors (some educational and others ideological) stand in the way of acquiring and applying a conception of the constitutive process. The constitutive process of decision making and the supporting expectations and beliefs people have about it seem to take form from existing situations that persist and flow into new patterns of accepted thinking and behavior. Functionally speaking, then, the constitutive process helps stabilize people's expectations about how society operates and how decisions are made. It is an innately conservative process, one that applies yesterday's solutions to today's problems.

A growing number of people are learning to see the constitutive process as a factor in natural resource management. Robert Healy and William Ascher, for example, in examining how new knowledge is used in natural resource policy making, noted that new knowledge seldom produces better policy.[72] They looked at how new information affects both ordinary and constitutive processes to explain their observation. New knowledge affects ordinary decision making by altering the intensity of the demands that people make, but it affects constitutive decision making by changing the arenas in which decisions are made, the power of political and bureaucratic institutions, and the skills required for participants to be effective in influencing policy in the relevant arenas. It also changes the values that gain or lose legitimacy and therefore influences the types of arguments that are seen as most legitimate and persuasive. International lawyers also use the concept of constitutive process to set up grounded, practical agreements and policy to manage the world's oceans, fisheries, national level resources, Antarctica, and outer space.[73] Aldo Leopold, considered the father of wildlife management, considered the constitutive process in his classic 1949 essay "The Land Ethic."[74] After looking at ethics in society over the last 3,000 years, he concluded by calling for new "rules" with regard to our relationship to nature, a new land ethic, in his words. Among his constitutive ideas: "When we see land as a community to which we belong, we may begin to use it with love and respect," and "a thing is right when it tends to preserve the integrity, stability, and beauty of the biotic community. It is wrong when it tends otherwise."[75]

Although it is clear that the constitutive process is problematic in greater Yellowstone, few people there see this basic challenge, its paramount importance, or how to manage it to advantage. This is evident in how leaders and others spend their time, what they talk about, and what they do. Not many could intelligently discuss questions concerning what "constitutive rules" are currently in place in greater Yellowstone. How is the constitutive decision process structured through existing institutional behavior? Do these rules and institutions permit necessary adaptations to serve common interests? How can we become more resilient as communities and as a nation, given the new demands that the changing world is making on us? More specifically, how can we adapt and stabilize new rules so that we can more easily address all the ordinary and political (or governance) problems? The real challenge in greater Yellowstone is to promote dialogue on these and related questions and to find workable answers, respectfully and democratically.

If we can make constitutive adjustments, then we will end up with fewer governance problems and fewer ordinary problems. We live in a democracy with national and state constitutions that call for creating and sustaining a commonwealth of human dignity for all Americans. To achieve this goal we need to adjust the constitutive process to adapt to changing circumstances.[76] The U.S. Constitution is intended to find justice, insure domestic tranquility, and prevent abuses of power; it also divides authority and control between the federal and state governments and among various branches of the national government and arranges institutions for representative democracy. What it does not do well is offer effective means to work out differences between the various levels of government. It does allow us to make the kinds of adjustments that are needed, however. We live in a time of unprecedented change in greater Yellowstone and throughout the world, with only more and greater change in sight. Given this context, constitutive issues are of the utmost importance.

Leaders in natural resource management usually are confronted with all three classes of problems. Because the stakes are high and the responsibilities enormous, they cannot rely on inadequate, outmoded, or conventional concepts, tools, and institutions. They cannot solve governance or constitutive problems using understanding and methods derived for use on ordinary problems. Today's problems require the utmost skill, the most modern understanding, and the best tools available. By way of example, Jerry Franklin, a leader in forest management, said that early in his life he simply lacked the "concepts and vocabulary" to understand the forest, its ecology, and management that later came to be the focus of his professional life.[77] But he added that though he had mastered ecological sciences, the concept of "policy" remained a "slippery" idea for him. Because Franklin's characterization is accurate for others as well as himself, the public is forced to depend on leaders who, though they have mastered the scientific or technical knowledge of their fields, may not yet have developed the knowledge, skills, judgment, sense of civic obligation, and scope to lead the community in making wise and far-sighted policy decisions about resource use.

Conclusions

Yellowstone has been described as a special place, a park, a region, and finally, an idea. It is also in its entirety an experiment, one that will, in a sense, never be finished, to preserve a parcel of nature, a dynamic landscape full of life. Through this experiment we have the chance to learn, if we are thoughtful enough, about ourselves and whether we are capable of creating a meaningful vision of nature, conserving an actual place, and sustaining ourselves in the process. It is problematic that this experiment is occurring in a landscape dominated by exploitive practices that are deeply entrenched culturally and institutionally and tied closely to notions of the frontier, the rights of independent people, private property, and unrestricted commerce. The preservationist aspect of this Yellowstone experiment stands in dramatic contrast to a widespread utilitarian management philosophy and the growing demands of local groups for power. Yet if current trends continue, the Yellowstone of the next generation will be very different from the one we know today.

If one thing seems clear today, it is that the future of Yellowstone, in every sense, is not sustainable under the trajectory and weight of our culture and its patterns of material consumption. To make it sustainable, one goal of leaders, according to the National Research Council, should be to "design strategies and institutions that can better integrate incomplete knowledge with experimental action into programs of adaptive management and learning."[78] Meeting this goal will require significant changes in current agency and professional norms and in broader governance and constitutive processes that are at the heart of our society. Despite the growing threats, the Greater Yellowstone Coordinating Committee and the various concerned NGOs and others can lead this transition. Much depends on what the region's leaders do in the near future.

The greater Yellowstone region has so much going for it. It is important here and around the world for many substantive and symbolic reasons. It is located in the wealthiest country on earth and is itself one of the wealthiest parts of that country. The ecosystem remains for now relatively healthy. It is an exciting arena full of challenges and active innovation. The region has a long history of enlightened management policy, and its leaders are part of a society committed to democracy, openness, and human dignity. Because of these factors, creating sustainability through transboundary or ecosystem management may be easier in greater Yellowstone than anywhere else on the planet. The burden of making this happen is on greater Yellowstone's leaders over the next few years. On their knowledge, skill, vision, power, and integrity depends the future of this beloved place, this historic park, this magnificent experiment, this world-changing idea.

2 Challenges Facing Greater Yellowstone

Greater Yellowstone faces many resource management challenges, some acute, some chronic. Some are ordinary and conventional, and some have to do with the less visible governance and constitutive dynamics, a mix that makes the leadership task especially difficult. It is a mistake to treat Yellowstone's challenges as only one set of ordinary problems, while overlooking or misconstruing the social interactions that lie at the heart of these problems. Because human process determines what happens to greater Yellowstone and its resources, leaders need a finely attuned functional understanding of the context within which problems and solutions take place—the decision-making process, their roles and responsibilities as leaders, and the problem-solving tasks facing them. They can then apply this understanding to the conventional modes of operating that presently dominate thinking and, it is hoped, move well beyond convention. This chapter is intended to help catalyze the functional understanding essential for improved leadership.

The challenges facing greater Yellowstone listed in this chapter represent the ideas of a wide range of researchers, government agencies, NGOs, popular writers, and others. It is helpful to look at what scholars and others think are the roles and responsibilities of the region's leaders, particularly the GYCC. In this chapter I suggest new ways of looking at the issues confronting these leaders by describing (1) a social process model by which they can understand these problems in context in more functional terms, (2) a decision process model that can help them ensure that decision making is carried out more effectively, and (3) the roles, responsibilities, and requirements of leadership that encourage leaders to be contextual. Leaders come and go, but the reward-incentive system and the organizational culture and structure under which they work are largely fixed and recurring patterns of leadership behavior. Systematic problem-solving tasks that leaders can perform are listed at the end of the chapter, beginning with analyzing the problems in relation to the goals that have been articulated. It is meant to stimulate further discussion, research, and action. Such a problem-oriented overview helps us understand the challenges facing leaders and what they might do about them.

The Greater Yellowstone Arena

Every policy system has an arena or zone of interaction. Structured by its institutional arrangements, an arena is an organized setting in which people interact as they pursue whatever they value. Greater Yellowstone can be thought of as such an arena, with the Greater Yellowstone Coordinating Committee as the highest-level, resident, federal leadership body addressing resource management problems. How the arena is organized to address ordinary, governance, and constitutive challenges should be of paramount interest to the GYCC and everyone else concerned about Yellowstone's future. Leaders' ability to detect and resolve problems is largely a function of how they help organize and participate in the arena. In turn, many other features of the arena affect the ability to lead, including the identity of the other participants, their demands, and the strategies they use to achieve the values they espouse.

The Yellowstone arena is a zone of interaction among a large number of people and organizations, typically with different interests, who use the interaction zone to their advantage and make demands on one another for favorable decision outcomes. Many features of the Yellowstone arena make it difficult to harmonize the diverse players, including the large size of greater Yellowstone, its history, the large cast of players, and the conflicting expectations and demands of these individuals. Beyond the inherent technical matters of making policy for the region, these practical and political dynamics make leadership roles especially challenging.

The greater Yellowstone arena came into existence as people began to share the expectation that other people and their decisions and practices would significantly affect them.[1] In the more than two hundred years since white Americans and Europeans first entered the region, human occupancy and use have increased dramatically. Political jurisdictions and patterns of government and private ownership have been established, making the arena increasingly complex. Extensive utilitarian use of natural resources has taken place in the form of mining, logging, livestock grazing, hunting, and agriculture. Recent decades have seen the emergence of an ecosystem approach to regional management. With these multiple uses and interested parties, the Yellowstone arena is highly fragmented, and there are very few mechanisms in place (or under contemplation) to address this fragmentation problem. Many entities compete for control: at least twenty-eight local, state, and federal government agencies are responsible and accountable for some part of greater Yellowstone, and two Indian nations and a large number of private individuals, businesses, and nongovernmental organizations are active in the region. The large number of participants makes it difficult to coordinate leadership, cooperate, and undertake joint action, especially for higher-order tasks.[2] One responsibility of leaders, such as the GYCC, is to organize the arena, including addressing the major problems caused by the fragmented authority structure. In considering how to organize the arena, many questions require consideration, such as whether structures should be centralized or decentralized, continuous or short-lived, organized or unorganized, specialized or general, and open to the public or closed.

One key aspect of this fragmented arena is how different groups define its spatial boundaries. The GYCC, whose authority and control are restricted to federal lands, calls the region the "Greater Yellowstone Area" (GYA) and includes only the national parks, parkways, wildlife refuges, and forests adjacent to Yellowstone National Park (see figure 1.1). Many environmental groups, academics, and other individuals, however, characterize the critical area in larger, more inclusive spatial terms, calling it the "Greater Yellowstone Ecosystem" (GYE). The GYE is delineated by the U.S. Geological Survey's map of watershed boundaries that encompass the two national parks and the surrounding mountain complex and intermountain basins.[3] Still others do not recognize the Yellowstone arena at all and think more parochially, in terms of county and state boundaries. Although there are multiple legitimate ways to circumscribe the area for the purposes of discussion and action, there has been little effort to find consensus on this fundamental point or to create a more workable arena.

The arena is structured in large part by government agencies, which are the most powerful organizations in the region and a key part of the institutional structure in our society. As the chief instrument of the "state" (either the federal or state government), the agencies "see like a state" and their bureaucratic ideology tends to simplify social and policy systems.[4] The arena as they define it functions in ways largely determined by the standard operating procedures of government agencies, including policies and regulations they have established, such as the National Environmental Policy Act. The rationality and constitutive rules for the functioning of the arena are obscure, but in practice a mixed system of formal relationships, precedent, and convenience has evolved that binds the governance apparatus, including intergovernmental relations, into some level of functioning. This current arrangement is the product of a long history of both accomplishment and conflict over contested power relations, that is, who should make decisions on what issues and when. In turn, this system greatly affects the structure, content, and dynamics of decision-making processes in the region. This context makes the GYCC's problem-solving efforts all the more difficult. The present political culture of the region amplifies these arrangements and their attendant policy difficulties.[5]

The dominant regional culture in the greater Yellowstone arena rests on a myth of the promise of America and faith in the wisdom of its early settlers.[6] People feel that some higher, invisible force accounts for their being "special." The culture promotes a sense of uniqueness, an American way of life, an American exceptionalism. There are strong sentiments on the notions of individual freedom, rights of individuals, self-determination, and independence. The culture promotes devotion, courage, and sacrifice to these beliefs. People are optimistic, faithful, and practical. The flag, the cowboy, and frontier symbols are tangible manifestations of this culture. It includes language that still invokes strong emotion. This myth, which glues the regional society together and animates it, is invoked in every situation to interpret and give meaning and coherence to events. In wolf reintroduction, for example, events are construed in terms of the regional myth. People are drawing on this myth when they believe that wolves kill too many elk and unduly threaten livestock and that bureaucratic mismanagement does not care for local people or their livelihoods.[7]

Historically, the regional culture has been dominated by local concerns. People are versatile in manipulating living conditions to meet dynamic lifestyle standards. A chronic and controversial issue has always been the proper scope of government—especially the federal government but also state and local bodies—in matters that affect private life. Responses by all levels of government to growing demands for change are often met with invectives from the conservative leaders who dominate local politics. Government action is typically seen as "creeping socialism or communism." At the same time, local people believe that they have a minimal dependence on government. But at the same time they assert a stubborn independence, westerners feel a strong sense of entitlement to government assistance. This regional cultural dynamic plays itself out in all natural resource management issues. It is a context that makes leaders' job difficult regardless of the issue at hand.

The way in which the arena, given the regional culture and other factors, is established and maintained by the GYCC and its agencies largely determines what kind of discourse (public as well as internal) is possible about Yellowstone's future. The real problem in the Yellowstone arena is not the conflicts that have continued around natural resource use since the park was set up, but the lack of an effective forum or means to resolve those conflicts. The structure and functioning of the arena influences the quality of the social and decision processes and the content of the management policies that are possible, including, for example, how power relationships and administrative arrangements are worked out,[8] who decides on problem-solving narratives and approaches, and who sets the boundaries on who has standing or access to decision making and on what issues.

A Conventional Survey of the Problems

As articulated by diverse authors and interests, greater Yellowstone's problems reveal the complex context in which leaders work. The GYCC leaders sometimes promote the ecosystem and transboundary concept in the policy debate about managing the area's resources, but this runs counter to the regional political culture. There have been times when public attention to government's behavior in managing resources has led to significant policy and administrative responses, but few of these have resulted in real solutions to the region's problems. In 1985, for example, the U.S. House of Representatives Committee on Interior and Insular Affairs held an oversight hearing on greater Yellowstone.[9] Congress was motivated to this action by powerful, vocal nongovernmental groups. At the time, the Greater Yellowstone Coalition and other people and organizations pressured Congress to carry out this hearing. About thirty-five people participated on ten panels, and a lot of additional testimony was received in written form. Testimony was heard from U.S. senators and congressmen, officials from the Departments of the Interior and Agriculture, and a wide array of special interest groups, from the International Union for the Conservation of Nature and Natural Resources, to the American Petroleum Association, to

Citizens for Multiple Use. Although this hearing revealed the diversity of competing interests in greater Yellowstone, the committee was unable to focus debate sharply or practically about management problems or solutions.

The Congressional Research Service (CRS) analyzed data submitted by federal and state agencies to the committee. Its 1987 report concluded that there was general agreement on the area's importance, but that the diversity of perspectives was manifest in the varied land management policies of the federal agencies, which showed a "thin mantle of inter-agency cooperation."[10] The report confirmed the obvious—that thousands of separate decisions produced overall management of the area. The wide range of management agencies, organizations, and interests, combined with a lack of shared data and lack of consensus about policy, created a complex context. Conflict was everywhere evident in the management debate then as it is now. The CRS report noted that one of the most serious deficiencies was an inadequate database. Information was absent or, when it did exist, "uneven." Data that were available were often not shared. Inconsistencies in attempts to coordinate across boundaries were also problematic. Moreover, "more of the same" kind of "cooperation" would not likely address the problems adequately. The report made suggestions to improve data collection and management, coordination of multiagency management efforts, and responses to other resource problems (for example, grizzly bear conservation).

Many authors have written about a wide range of problems in the greater Yellowstone arena in the years before and since the hearing and its report.[11] Most people's perceptions of the problems have focused on the management of ordinary problems. Not surprisingly, different observers see and describe problems differently. In one way or another, individually or collectively, these problems pose challenges to the future of greater Yellowstone and to its leaders.

Problems as Seen by Nongovernmental Environmental Organizations

The Greater Yellowstone Coalition (GYC) is a group of more than a hundred NGOs formed in 1983 to promote better policy and ecosystem management and "watchdog" government actions. It was the first to inventory problematic management issues in the arena.[12] The group issued a report in 1984 listing 88 problematic issues, and another report two years later identified 198 issues that GYC felt threatened the health of the region.[13] These included problems with wildlife management, wilderness and roadless lands, and timber and road building, among many others. The purpose of these documents was to bring these problems to the attention of agency officials, elected officials, and the public. In 1991 the GYC issued an "environmental profile" of greater Yellowstone in which it sought to define the ecosystem, its processes, components, and land management issues.[14] Human impacts were identified from hard rock mining, timber harvests, forest roads, oil and gas exploration and development, grazing, dispersed and developed recreation, agriculture and irrigation, and rural subdivisions. The profile included a brief case study of the GYCC and concluded that "today there remains no common approach to land management and no single entity empowered to assess the larger ecological ramifications of concurrent development activities within the Ecosystem." In 1994 the GYC produced a "blueprint for the future," a guide for sustaining greater Yellowstone, outlining the GYC's view of what form ecosystem management should take.[15] The GYC's newsletter, *Greater Yellowstone Report,* has covered some of these problems in depth over the years, but overall its focus has been on ordinary problems, with little explicit or systematic attention to governance and constitutive problems.[16] The NGO community as a whole has tended to define challenges and treat them as ordinary problems in their campaigns and allied tools (e.g., media releases, use of science and scientific management, involving outside authorities). They could instead target the underlying governance and constitutive processes that produce ordinary problems.

Other NGOs have also identified problems in the region. The Jackson Hole Conservation Alliance (originally the Jackson Hole Alliance for Responsible Planning), an activist group dedicated to responsible land stewardship in Jackson Hole, Wyoming, in the southern part of the ecosystem, has worked on oil and gas development, cattle grazing in Grand Teton National Park, overflights (flights over the national park by commercial and private aircraft), and many other issues. Like the GYC, the alliance's newsletters have covered these and other issues over the years, including the future of wildlife given the region's rapid development, the adequacy of county planning, bison management, land use and transportation, the loss of ranching, rapid growth, elk feed grounds, wildlife diseases, and highways and wildlife.[17] The Predator Conservation Alliance in Bozeman, Montana, takes on management of carnivores in greater Yellowstone and beyond. The Sierra Club and Earthjustice, also in Bozeman, have focused on a wide variety of issues and have been particularly active in grizzly bear conservation in recent years. These organizations advance their concerns with the public and through the courts by lobbying, education, and negotiations.

The Rocky Mountain Elk Foundation, also active in the Yellowstone arena, pursues elk habitat, development, feed ground issues, and diseases. Numerous smaller groups, such as local wildlife clubs, including the Jackson Hole Wildlife Foundation, focus on specific issues of local interest. The one thing all these organizations share is their focus on ordinary problems.

Problems Cited by Scholars, Researchers, Writers, and the Public

A number of reports and books by scholars, researchers, and writers have also characterized Yellowstone's problems. In 1986 Alston Chase's *Playing God in Yellowstone: The Destruction of America's First National Park,* a critique of the Park Service's management of the park, sparked widespread debate.[18] Chase criticized park managers for failing to base management on good science as he focused on elk and their impact on the northern range, wolf extermination and reintroduction, grizzly bear research and management, Fishing Bridge and Grant Village developments, the role of science in park management, and geothermal drilling.

Public attitudes and knowledge about ecosystem management in greater Yellowstone were surveyed in 1988 in a study that contacted more than three hundred residents, asking for their response to ninety-five questions.[19] Respondents were concerned about expanded government control and economic issues, fearing loss of local control over public and private land use practices. These are governance problems.

In 1989 Robert Keiter examined the relationship between the rise of the modern concept of ecosystems and its use in law, especially as it relates to greater Yellowstone.[20] In 1991 Keiter and Mark Boyce discussed the ecosystem management debate and its context and concluded that greater Yellowstone is seeing the rise of a new era of public land management, a new approach based on a fundamental realignment of human-nature relations, that this transition is not coming easily to the region or its leaders, and that, despite promising developments, "coordinated ecosystem-wide management is still more of a myth than a reality throughout much of the Greater Yellowstone region."[21]

Tim Palmer's 1991 book, *The Snake River: Window to the West,* argued that government agencies, local communities, and conservation groups have all failed to view the Snake River basin as a whole system, seeing it, rather, as isolated parts.[22] In his analysis, a holistic view was lacking in river basin management policy. "From people in charge in government agencies, from most elected officials, and from organizations that control the fate of the river, I found little vision that the future can be better than what we have now."

In 1994 Steven Minta and I looked at the problem of "defining problems" in greater Yellowstone.[23] Specifically, we looked at the difficulties of delineating the ecosystem's spatial and temporal dimensions and defining its management problems, and we discussed the implications of specifying "ecosystem management" as the solution to perceived problems. In a 1996 article Steve Primm and I asked, "What is the policy problem in Greater Yellowstone?"[24] Having identified scientific, economic, and bureaucratic problems, we noted that no one had made sense of threats overall but that conservationists believed that continuing present land use practices would further disrupt the ecological integrity of the region.

Paul Schullery in 1997 examined the cultural process of "inventing" Yellowstone as an ideal landscape of special meaning to the American public.[25] In his view, Yellowstone's history is largely a reflection of, or a reaction to, public values, attitudes, and actions or inactions.

I summarized the interconnected challenges in the region in a 1999 article, dividing them into three categories: contextual (rapid change, growth, pluralism, complexity, state-federal conflicts, and lack of a common or shared perspective), institutional (multiple organizations with overlapping authority and control and disparate mandates, uneven leadership, lack of creativity in problem solving, and resistance to change), and human (diverse perspectives and values and epistemological limitations).[26]

More recently Andrew Hanson and his colleagues looked at the ecological mechanisms linking nature reserves like Yellowstone National Park to surrounding lands and the ecological consequences of human demographic changes in the region.[27] They found that land use outside the reserves strongly affects ecological processes and biodiversity inside the reserves, translating into about a 5-percent (possibly as much as a 10-percent) loss of species. Species within Yellowstone are clearly at risk. They noted, too, that continuation of current trends will most likely cause a loss of ecosystem qualities.

What You See in Clear Water: Indians, Whites, and a Battle over Water in the American West, a 2000 book by Geoffrey O'Gara, explores water management on the Wind River Indian Reservation, Wyoming, in the southeastern part of the Yellowstone area.[28] For more than a hundred years, the reservation's Shoshone and Arapaho residents have been battling their white farmer neighbors over water rights. Although according to treaty rights, the Indians control the water within the reservation, the farmers have diverted the river water to irrigate their private lands. As a result of continuing irrigation, the Wind River has been dewatered in places, with dire consequences for the native fish population, among other harmful effects, and irrigation water returned to the river is highly polluted. This case, aspects of which are as yet unresolved, has been to the U.S. Supreme Court. Today it is being contested in contradictory management actions of tribal and Wyoming state governments.

A number of other analysts have looked at specific problems in the Yellowstone arena. In 2000 Christina Cromley examined an incident in which managers killed a "nuisance" grizzly bear in Grand Teton National Park, which led to a public outcry.[29] The incident opened a window on people's expectations and sparked a debate about how management decisions are made by officials. In 2002 she examined bison management in the Yellowstone region, centering on the controversial policies mandating killing of bison, because they are presumed to carry brucellosis, as they move out of Yellowstone National Park in winter.[30] She argued that the problem is not about bison and disease but about values, politics, and the structures of governance.

A 2000 study looked at the adequacy of existing management policy of the National Elk Refuge in Jackson Hole, Wyoming, focusing on the Jackson Hole elk herd.[31] The authors concluded that future decision making must avoid reducing problems to technical issues, being waylaid by inevitable conflict, falling back on problem-solving methods already proven to be unsuccessful, and using inadequate fixes for complex management and policy problems. The report also noted that leaders and staff who wanted to break out of old patterns of decision making had difficulty, given the powerful hold of bureaucratic and traditional ways of problem solving.

Murray Rutherford studied a unique effort by Bridger-Teton National Forest managers in the 1990s to incorporate the principles of ecosystem management into their forest planning process.[32] His study revealed a discouraging story about the adoption and implementation of ecosystem management in the Forest Service. Overall, the agency failed to incorporate the innovative or progressive aspects of the academic literature, neglecting to define key terms and emphasizing a limited group of natural values while neglecting others. He found that a lack of clarity about goals was the major problem, along with a lack of skills in collaborative problem solving and an inability of project teams to overcome many of the historical malfunctions in Forest Service decision making and bureaucracy. Learning was also very limited. He concluded that without significant institutional reform, future ecosystem management initiatives were likely to be ineffective.

In 2003 Gary Ferguson, in *Hawk's Rest: A Season in the Remote Heart of Yellowstone*, recounted his experiences in a remote area in Teton Wilderness just southeast of Yellowstone National Park.[33] Most irritating to him was that many uses of the wilderness were wrapped in the rhetoric of the American flag. Some local users, such as outfitters, misused rhetoric and symbolism as a license to despoil the wilderness. Many problems exist, he said, related to the fact that a complex mix of people uses the wilderness.

Even though most of these authors used the language of ordinary problems in describing the particular issues about which they wrote, they were all in fact talking about weaknesses in governance and constitutive processes.

Problems Identified by Government

A variety of problems has been identified by governmental agencies. Like the NGOs and others, the agencies exhibit a consistent pattern of defining and treating all challenges as though they were merely a collection of nearly independent ordinary problems. In 1997 the U.S. Geological Survey presented a proposal "linking science and management" in the Greater Yellowstone Area that listed "issues."[34] This proposal built on a 1996 information initiative involving the GYCC, Montana State University, and the University of Wyoming. Called "A National Spatial Data Infrastructure Information Center and Sharing of Geographic Information Systems Technology among Local, State, Federal Government within the Greater Yellowstone Area," this initiative was designed to instill cooperation among these groups as they moved toward integrated landscape management. The problems, though not specified, can be inferred from the proposal: (1) to assemble what is known and to fill in data gaps, (2) to measure change and to project where the changes are leading, (3) to evaluate the interrelations between human development and landscape systems, and (4) to put that information into integrated constructs that can be used to support land and resource managers in on-the-ground, real-time decisions.

In a closed meeting in Idaho in 1997 the GYCC identified more than thirty problems that needed management improvements: winter use assessments, protecting elk winter range and migration routes and patterns, large carnivores (that is, bears outside the recovery zone), recycling, recreational development capacity, visitor information ecosystemwide, travel management, hard rock mining impacts (air, water, noise), agricultural patterns (air quality), public expectations in the GYA, water quality and quantity, use fees, funding, noise, gateway communities (economics, ecosystems, coordination/cooperation), data collection management, partnership funding, overflights of the parks by commercial and private aircraft, ungulate numbers, appropriate recreational use (capacity, quotas, patterns), private recreational development (Big Sky ski area growth), implementation of special orders, shared facilities/resources, Beartooth Highway, jet ski use, livestock grazing, overall desired future condition (use), geothermal issues, wildlife disease, and red deer game farming.[35] Again in 2001 the GYCC listed problems it had been working on in recent years, including noxious weed management, wildlife studies and projects, soil and watershed management, Yellowstone cutthroat trout conservation efforts, whitebark pine restoration and management, land patterns, and recreation and visitor services.[36] The group's more current lists of issues and priorities cover lynx and wolverines, roadless and wilderness lands, fire management, coordinated information management and sharing of data, grizzly bear management, land patterns, noxious weeds, cutthroat trout, waterways, winter recreation, and whitebark pine.[37]

Yellowstone National Park identified eleven management "issues" in 2000: wolf restoration, lake trout, whirling disease, bison management, bison and brucellosis, bison management environmental impact statement, northern range controversy, grizzly bear recovery plan, grizzly bear conservation strategy, thermopiles (heat-loving microorganisms), and winter use.[38] In 2001 the park produced another list of controversial park issues, all of which spill over into the greater Yellowstone area, including bioprospecting, bison management, fisheries (lake trout and whirling disease), grizzly recovery and conservation strategy, northern range, winter use, and wolf restoration.[39]

It is typical that the problems identified by governmental entities are ordinary problems and that governance and constitutive problems are neglected. In two 1997 reports the General Accounting Office faulted both the Forest Service and Park Service for having weak decision-making processes.[40]

Problems with the Greater Yellowstone Coordinating Committee

A few authors have spoken directly about problems with the GYCC. For instance, according to Paul Schullery, the GYCC's work has consisted largely of "low-profile work . . . with routine administrative matters."[41] Don Barry, assistant secretary of Fish, Wildlife and Parks, U.S. Department of the Interior, noted at a GYCC meeting on July 21–22, 1998, that "he was unclear of just what the GYCC did/does."[42]

Environmental historian James Pritchard, who studied the role of science in park management, noted that the GYCC quietly carried out some federal-state coordination activities. These went largely unnoticed until 1990, when it undertook the "Vision" exercise to coordinate planning more formally.[43] The "Vision" (so called because of the title of the resulting draft document in 1990) called for government planning to coordinate management (lack of adequate coordination was one of the weaknesses pointed out by congressional hearings in 1985). The draft document unleashed a firestorm of controversy,[44] and several special interest groups, congressional delegations, and state legislatures pressured the agencies to water down the document and continue with status quo management. This disastrous and polarizing exercise led to a congressional Civil Service Subcommittee investigation, which concluded that "the Department of the Interior engaged in a politically motivated, underhanded operation to destroy the Draft Vision document because it was unacceptable to powerful moneyed commodity and special interest groups."[45] Perhaps high-level officials were being responsive to conservative, status quo–oriented regional and state politics. The final document was intended more to avoid offense than to offer any vision of an ideal future for the region, and since then the GYCC has reverted to "ordinary" management on a case-by-case basis, thus avoiding similar criticism. Pritchard's conclusion: "The politics were so highly charged that ideas about ecosystems or coordination of the simplest plans got lost in the dust."

Keiter and Boyce suggested that the GYCC's public involvement process is incapable of forging a mutually shared vision for the region.[46] The many decisions that individual agencies make are not being coordinated at present by the GYCC or any other group, which reveals "a troubling lack of consistency and an absence of defined priorities with the region's individual forests." Agency weaknesses, including inadequate coordination, make regionwide management problematic at best. These authors also noted a lack of coordination between state and federal agencies and concluded that more conflicts were likely to develop over resource issues and state-federal relations in the future. In addition, what happens on private lands outside the purview of the GYCC may well determine the fate of the Yellowstone region. Finally, they said that it is unknown whether the agencies are "seriously committed to ecosystem-based management in greater Yellowstone or whether interagency coordination is more myth than reality."

Finally, Mike Finley, then chairman of the GYCC, said publicly at the fall 1999 GYCC meeting that he viewed the GYCC as ineffective, a waste of time, and unneeded.[47] Informal, off-the-record comments by other GYCC members have revealed similar opinions of the committee, although in recent years members are more likely to support GYCC publicly. Lack of support from its own members brings into question its ability to implement the kinds of policies needed in the GYE, except over a very long time period and with the infusion of many more resources.

Functional Analysis of the Problems

It appears, then, that there are a great many challenges in managing greater Yellowstone's natural resources. How can leaders comprehend and deal with the overwhelming magnitude and complexity of such issues? Fortunately, methods exist that offer an explicit, comprehensive, systematic, rational, politically acceptable, and morally justifiable strategy. The first task that leaders can undertake is to understand thoroughly how decision making is accomplished and carry out a "net assessment" of the greater Yellowstone arena. This requires using concepts and methods specifically designed to accomplish this task. The second task is to make sure that all the relevant data have been collected and brought to bear in this assessment. This is a task in intelligence gathering, processing, dissemination, and appraisal. The data must be organized in useful ways so as not to overwhelm the analytic capabilities of leaders, their staffs, and organizations. The third task is to develop the leadership skills to lead the process successfully to common interest solutions. This requires skill and adaptability.

The social process for managing natural resources concerns people and what we value, how we interact, and how we establish and carry out practices that affect the environment and one another. If we want to understand management problems, we need to look not just at wildlife, habitats, housing developments, or biophysical scientific data, but also at ourselves and our practices and activities, deliberations, decisions, political wrangling, and other interactions. That is, we must understand in more functional terms how individual events, organizations, decisions, and other elements work in larger social and decisional processes.

In order to make valid observations, comparisons of interactions, and appraisals, we need universally applicable, purposeful concepts, standards, and language, using the same basic, functional terms, regardless of the issue at hand or the context. In short, we need a descriptive, yet analytic, framework that is up to the task. Functional thinking looks for connections, relationships, and systems properties in social and decision processes, data and connections that are frequently overlooked using only conventional means.[48] The functional approach explicitly draws on a comprehensive model of social process to guide attention to the interactions among people, organizations, and institutions.[49] The kind of social interaction and decision making that we decide to carry out not only determines what happens to ordinary issues, but it is also the very heart of our public order, that is, of our governance and constitutive processes. This view of management policy in greater Yellowstone is presently absent from most of the formal and informal deliberations among leaders as well as the public.

For example, a look at grizzly bear management in greater Yellowstone, from 1959 to the present Interagency Grizzly Bear Committee and its cooperating members, shows that nearly all research, education, and management has been done from a conventional standpoint confined to standard operating principles of bureaucracy. Put another way, the interagency bear committee has relied unreflexively on scientific management principles and the bureaucratic rules, roles, and regulations of the agencies to understand and address the bear management problem. History shows this square-peg-in-a-round-hole approach is extremely corrosive to interpersonal relations leading to more time in the courtroom, dueling scientific communities, and more. Overall, the committee's approach of "politicized science" and "scientized politics" has been detrimental to everyone involved and certainly led to a drawing down in trust in government and community social capital, both of which are needed to facilitate finding common ground in diverse communities. The focus has been limited to ordinary management concerns mostly. As a result, the social and decision processes—the "human dimensions," or much of what bear management is really about—remain largely invisible to officials and the public alike after almost half a century's research on bears and their management.[50] All the benefits that might have accrued from this larger, more insightful, functional understanding, such as creation of an ongoing forum for addressing and reducing conflicts, cooperative development of best practices for coexisting with bears, improved decision making about management, and interfacing more successfully with the public and local and state governments, have never materialized. There have been huge opportunity costs associated with the present bear management structure and its problem-solving and leadership style. The same can be said for wolf conservation, elk management, and many other issues.

Social Process

All the ordinary problems identified in the literature survey above can be viewed through a more functional lens. In this section I propose new ways for leaders to view those myriad problems in more systematic, functional terms. Doing so opens up new vistas of understanding. Here I briefly describe, first, the social process of greater Yellowstone, second, the decision process for managing natural resources, and third, leadership requirements in this context. In taking a more functional approach, we are shifting gears as we look deeper into the human aspect of greater Yellowstone, which brings us face-to-face with its governance and constitutive dimensions.

Social process is the ceaseless interaction of people as they carry out their lives and influence one another. Today we live in a world where the actions of people on the other side of the world affect us profoundly. The social process in greater Yellowstone is problematic in that it shows competing special interests striving to dominate the arena for their own advantage. But it also shows numerous attempts by diverse participants to improve the process to serve common interests—a promising trend.[51]

Social process can be "mapped" or analyzed in an effort to comprehend the social context in which all problems are embedded and which affects every detail.[52] A set of conceptual categories can help leaders to heighten their awareness of both the larger context and the details of particular situations. Every problematic setting, regardless of its subject matter, is made up of *participants* with *perspectives* interacting in particular *situations*. Participants employ whatever *values* or assets they have through different *strategies* to obtain desired value *outcomes*, which have long-term *effects* on society's institutions. Leaders can use this simple schematic to ask seven basic questions about any problem, including every one of those described above: Who participates in the process? With what perspectives? In which situations? Using which values as power bases? Manipulating them through which strategies? With what outcomes? And with what longer-term effects? This list, despite its brevity, reminds leaders to attend systematically and thoroughly to all the relevant contextual data that affects their problem-solving efforts. Each question opens a more extensive and intensive investigation into the particulars of any given context. Many leaders assume they have long since internalized the social process map on which their organizations function, but the discipline of creating and documenting a map usually demonstrates that one or more often critical corners of the map were, in fact, neglected.

The concept of values is a key one for thinking about how people interact. It is one of the functional concepts that can make problems operational and manageable as leaders try to understand the process involved in social interactions. Values can be thought of both as the things that people strive for in life and the assets they use to get them; they are the medium of exchange as well as the object of exchange. All human interactions consist of value interactions. Participants may "gain" certain values as they interact with others, or they may "lose" values. It is the interpersonal or transactional uses of values that leaders (in fact, anyone interested in solving policy problems) need to take into account. Some leading philosophers and other normative specialists recognize eight categories of values:[53]

Respect: freedom of choice, equality, and recognition
Power: making and influencing community decisions
Enlightenment: gathering, processing, and disseminating information
    and knowledge
Well-being: safety, health, and comfort
Wealth: production, distribution, and consumption of goods and services,
    control of resources
Skill: acquisition and exercise of capabilities in vocation, professions, and
    arts
Affection: intimacy, friendship, loyalty, positive sentiments
Rectitude: participation in forming and applying norms of responsible
    conduct

In healthy societies citizens enjoy a full range of values, a state that has also been called "a commonwealth of human dignity."[54] A healthy society is possible only when citizens enjoy a level of all eight values satisfactory to their needs.

We can ask a number of questions about the GYCC and its understanding of value dynamics in the region's social process. Does the GYCC perceive the values at its disposal as potential bases of power or influence? How does the GYCC use its values? Does the GYCC take a narrow, parochial view or a more comprehensive view of the value dynamics of which it is a part and which it helps to shape? What values does it seek to inspire in others? A key question is whether the GYCC uses its value assets, both symbolic and material, to establish and maintain a constitutive relationship among participants in the greater Yellowstone arena in an inclusive way? A constitutive relationship allows participants to address constitutive issues openly and adjust constitutive processes as needed. What are the consequences of the GYCC's behavior in ordinary, governance, and constitutive terms?

An initial social process map of greater Yellowstone (more research would bring detail to this description and possibly change it) would show leaders that the Yellowstone arena is becoming more pluralistic as new people and groups become involved in the self-organizing decision process. As more people are included or see themselves as part of the arena, the arena becomes more differentiated. In such an arena people tend to become more territorial, invoking a more active symbolic politics, as they seek to protect or expand their interests.[55] Perhaps this explains the increased politicization of many resource management issues in recent years. This is evident in the diversity of the problems now before the arena. The region's leaders can help make the arena work for this diverse body politic by arraying the values, both symbolically and in a real resource sense, in ways that empower people and provide access to the arena in order to facilitate finding common interest outcomes. It requires knowledge and skill to understand and lead the formation of an arena suitable to address ordinary, governance, and constitutive issues.

Although leaders have a responsibility to make a "contextual map" of each individual problem they face, here we might construct an initial, generalized map of the whole region, which would suggest how mapping can be done, starting with a list of participants. Participants in the greater Yellowstone arena include those individuals, groups, and organizations that are officially and unofficially, directly and indirectly, involved in the ongoing social and decision processes that determine the fate of the region. Because the arena is complex and covers a large geographic area, the group of participants is quite heterogeneous. It includes people outside the immediate region, too, although those who are most directly affected should be given priority. For each issue different participants come forward, and many of the major groups break down into smaller units, some of which play a role in decision making that differs from that of their parent organizations (for example, park rangers may have different perspectives, motivations, and constraints than the National Park Service). Public and private sector groups, corporations, and government entities operating within or close to the geographical area of the Yellowstone ecosystem must be considered participants. Residents living in or near the arena also participate at various levels in many different, overlapping arenas. Town governments, branches of state and national government, tribal governments, chambers of commerce, local businesses, groups promoting tourism and other activities, farming cooperatives, and religious organizations (and their national and international affiliates) all participate. Political parties and membership groups with political agendas, such as rural labor unions or indigenous peoples' political organizations, are also participants in greater Yellowstone's social process.

Every participant has a perspective, which may include a number of different, and indeed conflicting, ideas, feelings, and beliefs about a problem. Perspectives rest on basic beliefs (or myths) that are matters of consensus. Although social groups act with cohesion in many instances, participants are first and foremost individuals with their own patterns of attention, sentiment, interest, loyalty, and faith. Their identifications, or the way they see themselves as members of some aggregate or group, differ by individual and over time. People with similar perspectives form groups and loyalties and become the political "we/us" (as opposed to "they/them") in social process. One group that seems to be coalescing, for instance, shares a perspective in support of the greater Yellowstone ecosystem concept; other groups are forming that oppose the idea. Leaders—individuals at the forefront of building shared perspectives—must understand the dynamics of coalescing and diverging perspectives if they are to create an equitable system for resource management. They must understand that symbolic politics plays a role in the formation and articulation of perspectives and thus forms an element of the decision process. The failed "Vision" exercise by the GYCC in the early 1990s shows that it either did not understand these dynamics or did not know how to harness them. The ongoing grizzly bear and wolf management programs, as well as many other programs (including Jackson Hole elk management, feed ground management in western Wyoming, and bison management) also illustrate questionable understanding of the fundamental human value dynamics at play. Otherwise, the GYCC would not have set itself up for the criticism it received.

It is also vital that leaders seek to understand what values are held by the participants, how those values are being used, and which values are sought by whom and how, both symbolically and substantively. All participants, regardless of the value or mix of values they seek, want to improve the value "positions" in their lives; being better "positioned" improves the quality of their lives. In Yellowstone many participants seek more influence over the decision process. They want to insure that their perspectives and demands will dominate. People generally focus on the source of their greatest expected deprivation in terms of values, in other words, what they most fear losing. By empirically mapping each participant's demands (or claims) and his or her fears, leaders can assess which values are most important to each participant. One generalization is that all people want respect. Extensive discussions with people in and outside government show that most people in greater Yellowstone do not feel respected as individuals, as professionals, or as citizens.[56] Leading and making process improvements means understanding this fact and working to turn the situation around to cultivate more respectful relations among all involved people.

The strategies participants choose depend on their values and what they need to obtain their objectives and to mobilize the requisite support of others (by winning allies, by dividing opponents, or by neutralizing the opposition). Strategies may be educational, diplomatic, economic, or militant in nature. The range of strategies available to any participant is limited only by the intensity of commitment to the chosen objectives and by the strategies used by other participants, potential allies, and opponents. In assessing the use of strategies, it is important for leaders to pay special attention to the most powerful participants in the arena. For the GYCC, the most powerful participants in the arena are members of the House of Representatives and the Senate, appointed political figures in the relevant executive departments of the federal government, their immediate bosses, and key people in regional and local communities (e.g., elected and appointed officials in government, business leaders, and other influential people).

There are many outcomes, both symbolic and substantive, of the social and decision processes occurring in the Yellowstone arena. Outcomes are the short-term results of decisions, the specific products that emerge from the social process. Outcomes have different consequences for different participants. Considered functionally, some may gain in power, wealth, or respect, whereas others may lose out. Ideally, the reason to have a productive decision process is to maximize beneficial outcomes for all participants. Leaders' efforts to understand the arena and trends in decision making should thus include an analysis of the outcomes of past decisions and efforts to avoid inequities in future decisions. We can think of the social process in greater Yellowstone as one in which the arena is becoming more organized. Different participants are stepping forward to take a role in its organization and in its decision processes. One important initial outcome of this slowly organizing arena is the emergence of the greater Yellowstone ecosystem idea. This idea and the new arena that goes with it open up possibilities for more equitable decision making to serve the common good. Another outcome is that the ecosystem's management policy is now on the public agenda regionally, nationally, and internationally.

Outcomes, however, may also be negative for particular participants. While creation of an arena may direct attention to participation and communal process, it may also lead to structures dominated by extralocal organizations with disproportionate power in decision making and considerable distance from the outcomes of their decisions. For example, political appointees or high-level bureaucrats in regional and Washington, D.C., offices control key decisions affecting greater Yellowstone. Some local people also find it irksome that environmental organizations based on America's coastal fringes seek to play a role in greater Yellowstone in the Rocky Mountain heartland. It is also possible for structures to be dominated by parochial, local interests too. In either case, this limits access to the arena, constricts its democratic potential, and limits the opportunities to find common interest outcomes. For people whose livelihoods are based on the expectation of certain kinds of decision-making power over the use of resources, the outcome of the emerging Yellowstone arena may seem particularly unfair. In other words, opening up the arena to other participants will likely seem very threatening to those few who had previously had exclusive power. Therefore, leaders must strive to serve common interests and human dignity, which are ethically the bases for all decision processes, and to ground their decisions in reliable knowledge about differential outcomes and effects on different participants.

The effects of decisions must also be anticipated. Effects refer to the long-term changes in people's value positions and institutions. For instance, proponents of the Yellowstone ecosystem idea call for innovation—new perspectives and new practices in conservation and resource use. The diffusion of this demand already shows a mixed record in terms of institutional effects and changes. Some aspects of an ecosystem policy are being incorporated into management, whereas other parts are being restricted or rejected. The GYCC's behavior directly reflects this mixed diffusion/rejection process.

An examination of GYCC meeting minutes clearly shows this process in action (later chapters detail some of these). The minutes suggest how embedded policy preferences restrict a full, open examination of ecosystem management concepts or options. Given the large investment and institutional activity that is being marshaled in support of the idea of ecosystem and transboundary management policy by the nongovernmental and governmental sectors, it is clear that greater Yellowstone's decision process is perceived to have far-reaching consequences for many local, regional, and national actors. Given the current social process within which the Yellowstone arena is being organized, innovations and social change initiated under the rubric of "sustainability" will likely be partially used or incorporated and partially discounted or rejected by different participants in their operations. Leaders will need to monitor these trends. Their analysis of effects, as part of an effort to understand the region's social process, would thus consist of tracking changes over time in the status and satisfaction of participants. Data gathering for such an analysis might ask participants if they feel that the existing decision process has added to their power, wealth, enlightenment, skill, well-being, affection, respect, and rectitude. Participants' views should be solicited on the nature of particular projects, rules, and political processes that affect them. Their ideas about the future of greater Yellowstone, and their role in that future, should be sought and addressed. Such a program would need to monitor participants' views not only on the level of the entire greater Yellowstone's ecosystem, but also on the scale of selected smaller projects.[57] One problem, though, is that even if the overall governance and constitutive processes improve, if there are more and more people in the arena, the average "satisfaction" (e.g., respect, influence, sense of self in place, well-being, and so on) per person might go down and create perception that no progress has been made.

Decision Process

Understanding the context is a necessary preliminary to problem solving, but leaders also need a systematic way to understand the process of decision making itself and to help guide decision making toward effective and equitable ends. As people interact through social process, they make value-laden choices that have consequences for how resources are managed. The most useful way to conceive of policy making is that it is a process of making choices or decisions. It is this decision-making process that must be upgraded if we are to achieve sustainability. In a historic example, the early proponents of Yellowstone National Park and the U.S. Congress made decisions in the nineteenth century to create the world's first national park, decisions that affect us still. Today, many decisions, small and large, are being made that shape how the region is managed now and how it will be managed far into the future. We all participate in decision processes, directly or indirectly, in small or large ways depending on what we do, yet few people think about their actions as part of the decision process.

Understanding any one decision process in detail and comprehensively, in addition to the hundreds of other decision processes underway in greater Yellowstone, is a critical skill for leaders in resolving problems. It requires them to be "systems" thinkers. At present, some leaders are highly skilled at this, others are not. Knowledge about how decision making unfolds can help people watch for and avoid foreseeable pitfalls. In general terms, if leaders can configure social and decision processes so that they work for us all, then they will have a much better chance of finding common interest outcomes. Leaders who can do this are highly sought after.

Decision making about natural resources is part of larger social and decision processes wherein, despite their differences, people struggle to clarify, secure, and sustain their shared interests. In all the many smaller decision processes ongoing in greater Yellowstone at any one time, different participants want to use resources in different ways depending on their beliefs and value outlooks (for example, ranching, agriculture, biodiversity conservation, or protection of water quality). But we cannot assume that individuals are rational actors in these decision processes, performing cost-benefit analyses in every situation to determine what is in their own or the community's best interests. Management in greater Yellowstone is so contentious because many of the present decision processes are not carried out in ways that allow participants to find common ground.

All decision making includes or follows several interrelated activities, functions, or phases. Often leaders are unaware of these phases, and so they ignore, skip, or inappropriately attend to them. Decision making starts in *initiation*. Intelligence or information is gathered about the problem at hand and its context. Producing key information is essential. The next step is *estimation*, wherein information is debated and discussed and solutions recommended, advanced, and promoted. These preliminaries set the stage for *selection*, the decision itself, which is establishment of new rules or guidelines to solve the problem. The rules are then put into action through a program in *implementation*, which includes enforcement and dispute resolution. All of these activities are then monitored, evaluated, formally or informally, through *appraisal*. Finally, *termination* concludes the process when the rules are ended or succeeded by others. High standards exist for satisfactory fulfillment of each activity or function. It is extremely valuable for leaders to know about these phases, what happens or should happen in each, criteria by which to evaluate whether each has been successful, and the predictable weaknesses or pitfalls to anticipate.

Looking at each decision process activity in more detail can help us better understand what is involved in the decision process in general terms, regardless of the issue at hand. First, it is difficult to pin down exactly when a decision process concerning a particular issue is initiated. For instance, although the greater Yellowstone ecosystem idea itself has a number of beginnings, we might say that as a policy concept greater Yellowstone was launched though a collection of statements, press releases, and actions by some of the more powerful participants, particularly the academic and NGO communities, in the late 1970s and early 1980s. At some point, there was enough weight behind the idea of ecosystem management for it to gain legitimacy. For many people today, the idea has solid legitimacy, but for others it is still a discreditable notion. Eventually, more and more people picked the idea up, including other academics, managers, NGOs, citizens, and government officials.

Another example of initiation is the ongoing Jackson Hole Elk and Bison Environmental Impact Statement, which will establish guidelines for managing these two species for decades to come. It is difficult to identify a starting date for the present effort because this issue dates to establishment of the National Elk Refuge in 1912. There have been many decision processes since then focused on elk management, which has always been contentious and will likely remain so given the current decision-making structure as set by government. The standards for good initiation include timeliness, dependability, and creativity. Among common weaknesses during initiation are seizing on an overly simplified problem definition with simplistic objectives, domination by a single organization, assuming the program will benefit everyone, overlooking long-term consequences, and not balancing local and extralocal participation. Elk management, including the elk and bison environmental impact statement, has fallen short of recommended standards and shows common weakness in initiation efforts.[58]

The next phase, estimation, defines and in fact limits the possible solutions to a problem, and thus it is not a neutral activity. Rather, the kinds of knowledge produced about a problem reflect the underlying beliefs, values, assumptions, and biases of those involved. Greater Yellowstone always has numerous estimation, promotional, or recommending activities ongoing across a wide variety of natural resource issues. The production of working papers, management plans, scientific research, conferences, community meetings, press releases, high-profile debates in the media, and discussion all help participants to deliberate about and weigh the facts, dimensions, seriousness, causes, and directions of a problem or agenda. Competing interests seek to persuade voters or decision makers that their definition of a problem and thus their proposed solution are superior. For decision processes to result in more equitable outcomes, knowledge should be fully shared and openly debated by all participants. Insofar as possible, underlying assumptions and biases should be made explicit. Constructive debate can help to promote trust and cooperation. Common weakness in estimation include expert bias, expert opinions overruling the views of

those affected by decisions, exaggerated accounting of expected benefits and discounting of possible costs, and focusing on easily measured
or easily understood data and neglecting the "difficult" data. Hard
questions need to be asked in the estimation stage by leaders and citizens alike.

In the selection phase, a policy or management option is chosen
to resolve the problem at hand and guide future activities. Decisions
are made by individuals or governmental bodies with the appropriate authority and control to specify the guidelines, rules, or laws for
people's behavior. Authority is the legal mandate to set and enforce
rules, and control is the actual ability to influence the behavior of
participants in the arena. Within greater Yellowstone, for example,
many organizations—local, state, and national governmental bodies,
tribes, corporations, and NGOs—have different forms of authority and
control. To be most realistic and useful, the new policies or rules, when
they are set, should be specific about goals to be achieved, contingencies to be considered, sanctions to be applied, and resources required
to implement the rules. The reintroduction of wolves to Yellowstone in
1995, a highly contentious issue, was a good example of federal agencies making a tremendous effort to specify the goals, contingencies,
sanctions, and find adequate resources in the selection phase of this
ongoing management issue. Ever since the environmental impact
statement and the actual reintroduction, government has tried to address these issues in detail through its implementation efforts. This
has included following the Endangered Species Act (as government in-
terprets it) as well as being responsive to local and regional political
interests. Common pitfalls in the selection phase include weak government coordination, domination by strong government agencies,
agency rivalries blocking the achievement of program goals, and overcontrol of participants and beneficiaries.

Once decisions have been made, the decision process enters the implementation phase, during which chosen policies are put into place
and enforced through a program. In order for policies to have their
desired outcomes, implementation must meet certain standards. It
must be dependable, even-handed, realistic, and timely, and conflicts
over the implementation of policies must be resolved in ways that are
deemed fair by consensus of the participants. Weaknesses in implementation include intelligence failures and delays, wasting resources,
"benefit leakage" (where intended benefits fail to materialize on the
ground), falling victim to agency selfishness, lack of adequate coordination and appraisal, and inappropriate organizational arrangements.
In wolf reintroduction, implementation has experienced many of the

predictable problems, which have been detailed in newspaper articles, books, and court cases. The ongoing reluctance of the state of Wyoming to cooperate with the federal government and the governments of Idaho and Montana in producing an acceptable regionwide management plan is but one example of these implementation problems.

Even when decisions are made on the basis of the best information and with a high degree of consensus, they may not adequately address the problem in the ways anticipated. Initial conditions may also change, requiring shifts in policies. As time goes on, leaders and participants should evaluate how well the selected alternative has solved the original problem and, in larger terms, how well the overall decision process has served in achieving common interest outcomes. Of course, the success of implementation in meeting the needs of participants depends largely on how well the problem was estimated, goal clarity and specificity, and the point of view of the appraiser. In grizzly bear management today, for instance, the "delisting" (removing the species from protection under the Endangered Species Act) debate is based on appraising the status of bears, their habitat, threats, and adequacy of the present and future management. Some interests have appraised these to be adequate for delisting, whereas other interests have claimed that they are inadequate. Regardless of disagreements, monitoring and appraisal should be undertaken with the same caveats in mind as those that guide estimation activities. Common pitfalls in evaluation include insensitivity to criticism, failure to learn explicitly or systematically from experience, and ignoring the results of appraisal. Evaluation is a vitally important part of the decision process, but its effectiveness depends on whether the lessons learned from such exercises are returned to the participants and used to redirect and improve policy making and management.

Termination, the final activity of the decision process, occurs when the original problem is solved by the selected alternative, or when the participants conclude that the current decision process or its programs will not solve the problem, necessitating moving on to other approaches or issues. Often, if a decision process is successful, the institutions that were formed to address the problem remain, actively carrying out the duties assigned to them during the process. Thus, even when problems have effectively been solved, some part of the decision process remains active, allowing participants to remain involved in management on a routine, ongoing basis. Common weaknesses in termination include failure to anticipate opposition and manage it appropriately, seeing termination as a failure rather than adaptation, and dealing with people unfairly as the effects of termination become clear. Returning to the example of grizzly bear management, this complex decision process, covering decades, is now deep into the termination part, or the ending of one cycle of this process. Secretary of the Interior Gale Norton formally initiated action to delist in early 2006. She based her decision on claims of the U.S. Fish and Wildlife Service and allied groups that the grizzly population has recovered to a level of "health" that justifies state, rather than federal, management. Once bears are delisted, a new decision process will be initiated under state authority and control, and among the new decisions that will be made by the states is whether to hunt bears or not. Environmentalists, in contrast, have made counterclaims (as part of this termination activity) that the bear population still faces significant future threats from diminishing primary food supplies and habitat, in addition to ongoing mortality from hunters, cars, research, and other causes. Both sides claim to have substantial scientific data to back their positions, and both have tried to convince the public and leaders that their arguments are accurate and justified. Delisting, if it goes through, will certainly change how future decisions are made and by whom.

Decision processes, like social processes, can and should be "mapped" or analyzed by leaders as they seek to solve policy problems. Although decision processes occur at multiple levels throughout society, they follow similar patterns, and much research has gone into analyzing the different functions of decision making.[59] Although the decision process, regardless of the issue at hand, tends to move sequentially through phases, this is not always the case; different activities may overlap, take place simultaneously, backtrack, or get skipped. Leaders must pay close attention to decision processes and help guide them toward common ground outcomes.

Leaders

Mapping patterns of Yellowstone's leadership requires looking at the track record of the GYCC and other leaders in the region in a functional way. For example, we can determine how the GYCC organizes for decision making, how it organizes an arena for its operations, and what decision-making activities and standards it attends to. On this basis, we can infer what concepts of decision making these leaders are drawing on and what skills individual members manifest as they deliberate. By looking at the evidence of written materials and the empirical record of its performance and activities, we can begin to answer important questions about how well the GYCC is fulfilling its role and meeting its responsibilities. For example, is the GYCC clear about its goals and how it might shape common interest outcomes in decision-process terms to achieve its goals? How does the group participate in the ongoing social and decision process in the greater Yellowstone arena? What does it focus its attention on? How does it spend its time talking and acting? What decision process activities does the group actually undertake, and which ones are overlooked, ignored, or avoided? What standards does it seek in decision making? What efforts do leaders make to avoid known decision process weaknesses? Overall, how does the GYCC conduct itself?

Ideally, leaders should be viewing greater Yellowstone through the lens of social process mapping. This can give them a systematic way to organize their attention and remind them of the full range of contextual variables they need to consider, thus making problems and their resolution more manageable. Similarly, viewing problems through the lens of decision process mapping can give leaders knowledge of the activities they are caught up in, standards by which to measure their performance, and finally, common weaknesses to avoid. These decision-making concepts are not academic abstractions but extremely powerful, practical tools that can enormously strengthen leaders' fitness and competence.

On the surface, leadership is about setting goals, mobilizing support, planning, hiring staff and co-workers, issuing orders, assigning roles, tracking progress, and much more. To be successful, leaders must perform all these tasks. Looking at these tasks in more functional terms, leadership is about guiding social and decision processes and adapting them to the realities of the context of operations, not only the local context but the national and global contexts as well. The job of leaders is to help the body politic get these processes "right," given our goals of democracy and sustainability. This requires a particular set of leadership skills.[60]

The GYCC's basic objective in the decision process, functionally speaking, should be to integrate competing and divisive claims and counterclaims about how greater Yellowstone's resources should be used and who should decide. The process also needs to address perceptions and serve common interests.[61] Leadership in such decision-making situations requires clarity of standpoint. Leaders must distinguish their roles and responsibilities, their own and other participants' perspectives, and they must figure out how these might be integrated practically into a common interest outcome. What appears to disputants as a direct clash of interests may appear to the objective leader or systematic observer as an opportunity to bring people together and share benefits through new structures of cooperation.

Leaders must be creative in their search for integrative solutions to decision making at all levels.

Some natural resource policy experts have documented a growing crisis in policy making throughout the West.[62] Clearly, both social and decision processes in the Yellowstone region are problematic and growing more complex. Good leadership can guide these processes toward the overriding goals of democracy, human dignity, and sustainability. Leaders thus must cultivate the best conceptual tools, analytic skills, and knowledge because there is a great deal at stake for them and for us all. Successfully leading a transition toward sustainability requires shifting to a new equilibrium between resources and society. Leaders must not only help society make this transition, but they may also need to "retool" themselves to do so.

Problem-Solving Tasks

In addition to understanding the social context within which problems take place and the functions of decision making, leaders can greatly benefit from developing a strategy for analyzing problems fully and inventing solutions.[63] The myriad problems surveyed earlier in this chapter can be tackled in a way that makes sense of them and provides a much greater chance of devising effective and efficient solutions. This kind of problem orientation puts the decision process into a framework that focuses on the rationality of problem solving as well as on the content of the problem at hand, while also being sensitive to the political and moral aspects of the context.[64] It gives leaders a systematic set of tasks to undertake as they set out to solve a problem. In problem orientation, the problems at hand must first be specified in relation to the goals that people seek, thus permitting a clearer definition of the problems than is otherwise possible. Historic trends must then be described to see if events and decision making are moving toward or away from the specified goals. Next, factors or conditions that have influenced or caused these trends must be determined. When past trends and conditions are adequately known, projections of future trends are possible. Finally, after these four tasks have been completed and the necessary information assembled, alternative courses of action for achieving the stated goals can be invented; evaluated according to their effectiveness, efficiency, and equitability in solving the problem; and selected. The first four tasks are necessary to define a problem adequately, and the last can only be figured out using the information from the first four. If these five tasks are carried out comprehensively, yet selectively and realistically, a practical solution to the problems can most likely be found.

Given the intensity of the public and professional debate about managing greater Yellowstone's resources, we must apply even more analytical reflection about the nature of the problems. Participants in such debates are often so eager to promote their own favorite solutions that they neglect to ask or analyze carefully and systematically the nature and scope of the problems they are trying to solve in relation to the goals they want to achieve.

Many authors have emphasized the absolute necessity of defining problems realistically before promoting, selecting, and implementing solutions. Political scientist Janet Weiss made clear the importance of problem definition: "A problem definition at the outset of the policy process has implications for later stages: which kinds of evidence bear on the problem, which solutions are considered effective and feasible, who participates in the decision process, how policies are implemented, and by which criteria policies are assessed. . . . At whatever stage a new problem definition gains significant support, it shapes the ensuing action. It legitimates some solutions rather than others, invites participation by some political actors and devalues the involvement of others, focuses attention on some indicators of success and consigns others to the scrap heap of the irrelevant."[65]

Specifying Possible Policy Goals

Greater Yellowstone is one of the largest, essentially intact ecosystems in the temperate zones of the earth, and the richest, most nearly intact complex of wildlife and wilderness in the lower forty-eight states.[66] It supports local economies based on recreation, tourism, resource extraction, and ranching. Consequently, it is crucial that leaders design management appropriate to this unique region and ecosystem. But management of greater Yellowstone has broader implications. Yellowstone has enormous national and international importance, so its management policy standards act as precedents for other regions worldwide. If its leaders can bring about changes that will satisfy peoples' demands and lessen conflict, possibly through ecosystem or transboundary management, such policies are likely to be emulated elsewhere in the world.

One indication of the vast potential and significance of greater Yellowstone is the variety of goals that could realistically be contemplated for the region. In what other area of the continental United States, for example, would it be feasible to maintain a large population of grizzly bears for the long term? Possible goals for greater Yellowstone might include sustaining economic growth or sustaining development; maintaining the area's unique hydrothermal features; maintaining and developing extractive industries, such as logging, mining, and oil and natural gas production; preserving wilderness areas; continuing historic ways of life, such as ranching, outfitting, and guiding; maintaining open space; developing recreational opportunities; preserving viable populations of migratory ungulates and large carnivores; preserving a large proportion of the region's biodiversity in general; promoting patentable biotechnical genetic material; or simply leaving nature alone to proceed on its own course as free of humans as possible. Many of these goals are currently being promoted for the region.

Unfortunately, all of these goals cannot be achieved in the same place at the same time. Although there are certainly some win-win solutions available to achieve multiple goals in greater Yellowstone, some goals will still conflict in practice. Readily apparent conflicts exist not only between the goals of different communities and identity groups within the local region, but also between the goals of local, national, or international interests, and between goals for the immediate future and those for the longer term.[67] In some cases conflicting goals call for dramatically different management strategies.

How should greater Yellowstone's leaders, specifically the GYCC, allocate priority among conflicting management goals, and how should they help participants pursue the selected goals? In policy-oriented language, how can leaders help clarify and secure the common interest concerning management of the region? Much of the literature on this matter simply promotes personal and special interest conceptions of the common interest. But the views of any subset of interests in the region may not capture a true common interest; ultimately, clarifying and securing common interests is a task of governance.[68] In fact, the principal function of democratic government is to discern and pursue the common interest from among the many conflicting short- and long-term goals espoused by "communities of location" and "communities of interest." An appropriate higher-order goal for greater Yellowstone's leaders, therefore, might be to develop institutions and processes of governance that are capable of this task, so that the policies that are chosen will enjoy broad-based and enduring public support.[69] In order to cope with the uncertainty and unpredictable surprises that confound natural resource policy making, leaders should ensure that this process of clarifying and securing common interests is continuous and adaptive and that it incorporates margins of error (in case participants' initial choices turn out to be wrong).[70] Moreover, to be effective, governance must operate (or at least have influence) at geographical and temporal scales that encompass the major ecological and social parameters that ultimately will determine the outcomes of management decisions. Otherwise, decisions will be frustrated by the operation of variables over which managers have no influence.

Looking Back at Key Ecological and Sociopolitical Trends

A few trends stand out as especially important for leaders to consider in managing the region. First, socioeconomic circumstances in greater Yellowstone are rapidly changing. Increases in tourist visitation coupled with higher numbers of permanent residents have driven expansion of the service sector economy. Together with reductions in allowable timber harvests from the national forests, this means that the relative importance to local economies of logging has significantly declined.[71] Other extractive industries also face threats. Increased human presence has had additional impacts. Some economists argue that tourism has pushed ecological resources to the limits of sustainability.[72] Newly constructed housing for incoming residents is threatening ecologically important lower-elevation lands and socially valued open space.[73] In addition, new residents' values with regard to the environment often differ substantially from those of long-term residents.[74]

Socioeconomic change and increasing demands on finite resources have also led to heightened conflict. Environmentalists argue with natural gas industry representatives over drilling on national forest and Bureau of Land Management lands. Climbers argue with wilderness advocates over the use of fixed climbing aids in areas designated as wilderness. Business owners and developers conflict with conservationists and preservationists. Ranchers argue with conservationists over the reintroduction of wolves. Policy makers caught in the middle of these conflicts know that they will face criticism for any decision they make, and if they make no decision they will be criticized for inaction.

Ecologically, an important trend has been toward reduced biodiversity.[75] Although managers have had some success in expanding populations of such prominent species as elk and wolves, many less well-known species are declining. Even the recovery of a prominent species like the grizzly bear is disputed and its future uncertain.[76] Eleven plant species, six fish, one reptile, twenty birds, and eighteen mammals were listed at risk in greater Yellowstone in a 1989 report; since that time, more species have become threatened.[77] Well over a hundred invasive species (alien colonizers) have infested the region, causing significant damage. Perhaps the most significant trend from a management perspective is the ever-increasing evidence of the impacts of the region's people on its ecological systems. Repeatedly, management decisions made on the basis of short-term needs, single species, or other narrow foci have been frustrated or have produced unexpected consequences because of unforeseen linkages across space and time, often caused or affected by humans. For example, fire suppression practices designed to control short-term fire damage contributed to the long-term buildup of fuels that caused major fires in 1988 that were beyond the capacity of the interagency effort to contain them. Similarly, closing the garbage dumps in Yellowstone National Park in the 1970s led to increased human-bear conflicts as grizzly bears moved beyond the boundaries of the park in their desperation to find food. There are many other examples of human effects on the ecosystem: water use in areas outside Yellowstone National Park threatening geothermal features in the park; housing subdivisions interfering with big game migrations; backcountry skiing and snowmobiling stressing wildlife on winter range.

This mounting evidence of the interrelatedness of social and ecological systems has fostered a demand, expressed across a wide variety of issues, for management on broader spatial and temporal scales. Calls for ecosystem-level and transboundary management have been made by analysts discussing grizzly bears,[78] fire,[79] ungulates,[80] wolves,[81] bald eagles, trumpeter swans, peregrine falcons,[82] geothermal features,[83] geoclimatic characteristics,[84] oil and gas development,[85] and regional economies.[86] The need for, at the very least, better coordination among agencies is almost universally acknowledged.[87] Some attempts have been made to respond to the demand. The Interagency Grizzly Bear Committee, the Interagency Fire Control Center, the Greater Yellowstone Coordinating Committee's aggregation of National Park and National Forest Management Plans, and the GYCC's "Vision" exercise all represent efforts to coordinate decision making. To date, none has been a resounding success or a model for future emulation. In particular, the widely negative reaction to the GYCC's "Vision" exercise makes it clear that efforts to date to clarify and secure the common interest have fallen well short of recommended standards.[88]

Analyzing Conditions under Which Trends Have Taken Place

The major conditions and causal factors underlying these trends are well recognized. Throughout the literature covering the Yellowstone region, reference is made to the large number of management agencies (at least twenty-eight) with conflicting jurisdictions and mandates and the resulting highly fractured authority and control over the region. Coordination in such a situation is extremely difficult. Fragmented management has also produced an information base that, although better in recent years, remains inadequate for politically relevant or broadly justifiable decision making for the whole region. Ecological and social data are patchy at best and in too many cases nonexistent. Data that do exist are often poorly communicated within and among agencies and between the agencies and the public.

Multiple jurisdictions and mandates not only hamper management of ecological systems but also make it difficult for leaders to coordinate human systems in an inclusive fashion. Extensive housing development in any part of the region, for example, affects social and ecological conditions over a much larger region because of transportation demands, water pollution, and other broad-scale impacts.[89] When adjoining jurisdictions take very different approaches to housing and commercial development, the task of designing and implementing an effective regional growth management strategy becomes impossible. Tourism and recreation management faces similar challenges.

Other important conditions include the demographic profile of the nation's population and the technological characteristics of modern work. As the baby boom generation becomes wealthier and retires, tourism and recreational demands will continue to increase, and new residents will continue to move into the beautiful and desirable Yellowstone area. Further pressure will come from workers who, because of modern technology, are free to choose where to live and conduct their work and who seek quality-of-life characteristics, such as aesthetics and recreational opportunities.[90] These conditions are causing similar upsurges in population and use throughout the entire Yellowstone to Yukon region.[91]

The conditions outlined here can and do by themselves lead to escalated conflict, but the conflict is intensified because there is no adequate forum (outside traditional political and bureaucratic arenas) for disputes to be aired and resolved at either small or large scales. Some degree of conflict can be helpful to leaders by providing feedback that policies need to be adjusted or abandoned.[92] But when there is no meaningful arena or opportunity for differing values and perspectives to be heard fairly, conflict festers and can develop into social crisis. The controversy about protection of the spotted owl in the Pacific Northwest was a classic example of this.[93] Without alternative forums, disputes that center on value and interest differences and public policy matters may end up in court, where proceedings are adversarial and the focus is on issues between the immediate parties rather than on the broader public interest.

Projecting Likely Scenarios for the Future

If these conditions remain unchanged, a worst-case projection of current trends suggests a bleak future for greater Yellowstone. Ecological and social amenities will be degraded by poorly regulated pressure from tourists and new residents. Biodiversity will be impoverished. Lands and resources that are ecologically and socially interconnected will continue to be managed under fragmented, short-term, and perhaps contradictory policies. Widespread oil and gas development on industrial scales will alter landscapes, reduce wildlife, and harm local communities far into the future, especially in the southern portion of greater Yellowstone. Conflict and frustration will rise until trust and other elements of social capital severely erode. Trust in government will decline significantly. Dialogue among interest groups will become more difficult, management problems will continue unresolved, and eventually a major political crisis could erupt.

Bill Romme, a plant and fire ecologist, summed up the most likely future for greater Yellowstone under present trends and conditions.[94] He doubted whether we would experience a crisis of "sufficient impact and clarity to precipitate a revolution away from traditional ideas of resource management. Erosion of biological diversity, loss of open spaces, diminishment of air quality, and most of the other degradations that we are concerned about, usually occur gradually, by small, barely perceptible increments. . . . My darkest fear is not [that] the earth's ecological systems will one day collapse; but that they will be progressively diminished in richness and in beauty, and that the mass of humanity will not even realize what we have lost."

Evaluating and Selecting Possible Alternatives

The scenario described above is not inevitable. Addressing underlying conditions can change projected future trends. But time is short and harmful trends are accelerating in greater Yellowstone. Among the alternatives suggested in the literature are proposals to merge all five national forests into one, to create a superagency with authority and control over all the federal lands in the region, to improve existing interagency coordinating committees, or to develop new interagency and agency-private partnerships.[95] Other possible alternatives include

privatizing all federal lands or devolving management to state and local entities. These and many other proposals are on the table to address
small-scale, short-term problems as well as large-scale, long-term problems. Because there is currently so little trust among participants in
the overall policy debate, few of these alternatives are likely to gain
broad support.

Clearly, there is no magic bullet or simple resolution to the growing
interconnected problems. But greater Yellowstone's leaders have a duty
to orient fully to the many problems facing the region and to make a
concerted, renewed effort to help participants find their shared interests and sustain this unique ecosystem. Ecosystem management and
transboundary initiatives are the most promising alternatives forward
at the present time to help address the interconnected problems. They
offer new, flexible ways for people to interact and explore new management policy goals, and a vehicle for trying new approaches on the
ground, for learning, and for adapting.

Conclusions

Thousands of pages have been written about greater Yellowstone's
management and policy problems. Although the problems are typically described in conventional terms and include a long list of ordinary problems, many writers also speak to the underlying governance
and constitutive challenges. In one way or another they all address decision making, specifically, perceived weaknesses in decision making
and leadership. Despite its claim that it does not make decisions, the
GYCC is in a unique and authoritative position among greater Yellowstone's leaders and organizations to address this multitude of problems on all levels, to use the considerable resources at its disposal to
bring people together in a shared vision of the future, to build exemplars of cooperation and coordination, and to demonstrate real leadership. In order to accomplish these goals, and in order to improve
management and policy, however, the GYCC must first define the problems it faces in a realistic and functional way. The social and decision
process models described briefly in this chapter can be most useful to
leaders in characterizing the context of problems as well as the functions of decision making that must ultimately be carried out.

Leaders must recognize that, stated most simply and broadly, management and policy making is the process by which *people* seek *values*
through *institutions* using *resources*.[96] This is the never-ending process
of natural resource management policy. This functional understanding
of the activities of the arena is a critical first step. The GYCC's leader-

ship task is not only to participate in social and decision processes but also to develop new levels of effectiveness in serving common interests. The group can become more effective by guiding participants through the systematic tasks of defining problems in relation to goals; analyzing the trends, conditions, and projections of those problems; and evaluating and choosing better alternatives.

The Yellowstone ecosystem is interconnected biophysically, economically, and socially. This situation spawns many challenges—ordinary, governance, constitutive—as people dispute which values will dominate and whose interests will be met. The GYCC has before it a rare opportunity to create a policy model for managing natural resources that can revitalize conservation efforts for large-scale systems everywhere.

3 Leaders—Problem Solving

There are many leaders in greater Yellowstone, but few have as much influence over management policy as the Greater Yellowstone Coordinating Committee, composed as it is of the leaders of the national parks, forests, and wildlife refuges in the arena. Because of its visibility, mandate, and resources, this organization's behavior speaks directly to the status of leadership in government throughout the region. But are these leaders explicitly skilled in integrated problem solving, do they have the skills that would be most effective in helping the region transition toward sustainability?[1] Problem solving has been well studied,[2] and it is known that some ways to solve problems are far better than others, requiring users to be both substantively and procedurally rational, to be fully contextual, and to use multiple methods.[3] Skill in problem solving, in turn, requires leaders to be good at cooperating with others and demonstrating what needs to be achieved. A clear, long-term pattern in the GYCC's handling of these three key skills—problem solving, cooperating, and demonstrating—emerges from a review of its history.

In looking at the problem-solving approach used by the GYCC, I will consider the committee's composition and history, survey goals the group has set for itself, and analyze its strategy and behavior in engaging the problems that it takes on, and I will also compare the problem-solving strategies being used against best practice standards. In chapter 4, I will look at how the GYCC goes about the other vital leadership tasks of cooperating and demonstrating, and in chapter 5, I will examine factors that influence leaders as they carry out these three tasks. Moving a large, complex system to new levels of operation is difficult even under the best of circumstances, and to be successful, leaders must address many interrelated challenges rationally, practically, and morally. This requires integrated problem solving and "transformational" leadership,[4] distinguished by skills for effecting change, innovation, and entrepreneurship.[5]

Composition of the Committee

The committee's size and membership have varied over the years. Its members are "unit managers," the highest ranking members of, originally, the National Park Service and National Forest Service units in greater Yellowstone. In the 1990s the committee was expanded to include a unit manager from the Fish and Wildlife Service. Even though the regional director of the Intermountain Region of the National Park Service and the regional forester from the Rocky Mountain Region of the Forest Service are listed as members, they seldom attend meetings.

In 2000, for example, the nine members of GYCC were six forest supervisors, two park superintendents, and one refuge manager. These nine people (four women and five men) had an average of 27.8 years (range 20 to 38) of professional experience (collectively, 241 years of experience). They had served on the committee for an average of 6.2 years (range 2 to 10 years) and had been in the greater Yellowstone arena an average of 9.3 years (range 4.5 to 15 years). Together, they controlled $113.5 million dollars of annual budgets and 1,376 full-time employees or equivalents. Six held bachelor's degrees, and three held master's degrees. The committee's first coordinator had twenty-five years' experience and a professional profile similar to that of the members.

In 2003 the committee changed, with four new members replacing retiring members and a new coordinator taking over (table 3.1). The turnover in membership was relatively great between 2000 and 2003. The nine members in 2003 had an average of 26.7 years (range 24 to 34) of professional experience (together, 241 years' experience). They had been on the GYCC for an average of 5.1 years (range 2 to 12 years) and in greater Yellowstone an average of 7.8 years (range 1 to 15 years). Together, they controlled $109.7 million dollars and 1,287 full-time employees or equivalents. The annual budgets for individual units ranged from $150,000 to $22.7 million (mean $11 million). Staff averaged 128.7 people (range 1 to 271) per unit. Five members were women. Five had bachelor's degrees, and four had master's degrees. The second coordinator had twenty-five years' experience, a master's

Table 3.1. Composition of the Greater Yellowstone Coordinating Committee and Members' Experience, 2003

Management Unit Represented	Years on GYCC	Year of Experience in Greater Yellowstone	Bachelor's Degree	Advanced Degree	Annual Budget[a] and Number of Employees	Other Relevant Experience
Shoshone National Forest	8	8	B.S., natural resources, Humboldt State, Calif.		$8.3 million 90 PFTs	25 years with Forest Service
Gallatin National Forest	2	2	B.S., recreation and environmental education, Oregon State Univ.	M.S., forest management, Univ. of Wash.	$11.7 million 158 PFTs	24 years with Forest Service
Custer National Forest	10	10	B.A., English literature, Douglas College, N.J.	M.A., anthropology and archaeology, Univ. of Arizona	$6.2 million 80 PFTs	25 years with Forest Service; 3 years with National Park Service
Beaverhead-Deerlodge National Forest	1	1	B.S., geology, Portland State Univ., Ore.	M.S., natural resource management, Colorado State Univ.	$16.2 million 193 PFTs	26 years with Forest Service

Bridger-Teton National Forest	5	10	B.S., biological resources and secondary education, Colorado State Univ		$12 million 120 PFTs	28 years with Forest Service and Bureau of Land Management
Caribou-Targhee National Forest	10	12	B.S., forestry, Univ. of Idaho	M.S., range science, Univ. of Idaho	$22 million 225 PFTs	34 years with Forest Service
Yellowstone National Park	1.5	1.5	B.A., American history, Univ. of Western Florida		$22.7 million 271 PFTs	25 years with National Park Service
Grand Teton National Park	1.5	11	B.S., natural resource management, Univ. of Arizona		$9.2 million 137 PFTs	28 years with National Park Service
National Elk Refuge, National Wildlife Reserve	7	15	B.S., wildlife biology Colorado State Univ.		$1.2 million 13 PFTs	26 years with Fish and Wildlife Service
Executive coordinator for GYCC	1.5	30	B.S., fish and wildlife management, Montana State Univ.	M.S., fish and wildlife management, Mont. State Univ.	$150,000, 0 PFTs	27 years with Forest Service and NGOs

[a]Base budgets, but not total funds managed. PFT is permanent full-time equivalents.

degree, and a profile similar to the members. Grand Teton National Park had another turnover in its superintendent position in 2004, so a new representative joined the GYCC at that time.

Implications

What can we conclude about the committee's composition and experience? More important, is this the best structure for transitioning toward sustainability, given the range of challenges? Several things are clear. First, the GYCC's composition consists of people with great experience, especially in federal agency, bureaucratic settings. All the unit managers successfully came up through the ranks of their agencies after years on the job. It is reasonable to assume that the members are as qualified, if not more so, than any comparable group of administrator-leaders in the world.

Second, because the members are responsible for their own management units first and foremost, their perspectives are shaped by the needs of their units. This is what they were hired to do, considerable authority and control was given to them for this purpose, and the reward and incentive system they live under promotes this focus of attention. Membership in the GYCC is a secondary interest, although perhaps one that is growing in importance. The members share, more or less, a value orientation and a frame of reference, which includes "sanctioned patterns prescribing the approved way of doing things and the established goals of that body."[6] They are firmly anchored in the status quo and spend most of their time dealing with the many day-to-day, ordinary problems of their units. Their time and attention is taken up by the daily responsibilities related to their parks, forests, and refuges. Moving beyond this "home unit" imperative to consider the greater region and the need to cooperate in strategic ways could be a stretch for some members.

Third, experience has tested members in different ways. They are all clearly successful administrators, a role that Burns calls "transactional" or "maintenance" leadership. Although their leadership experience and skills were predominantly developed in similar ways, in the context of bureaucracies and administrative requirements, this does not mean that they all think and act alike. Administrative leaders lead by making small improvements at the margin of their bureaucracies. New practices come slowly, and deep changes are rare. Transactional leaders are responsive to their staffs' interests and provide rewards for services successfully accomplished according to the established rules and norms.[7] In organizational contexts such as the one in greater

Yellowstone, where politics is paramount, organizational cultures of power and role tend to be rigid and leaders are risk aversive. They play it safe. In contrast, in situations that are not as problematic and where political considerations are less important, the organizational culture can be more task oriented, people centered, and organic, and leaders can take more risks. There are many gradations in leadership styles on the committee and throughout greater Yellowstone.

Finally, coordinating across management units is a relatively recent task, given the more than one hundred years of agency history in the region. It requires leaders to meet complex, higher-order tasks of strategic leadership and to engage in deep policy reflection. Yet both the higher-order and deeper reflective tasks must be undertaken in the same inherently conservative, status-quo-maintaining, bureaucratic setting and political context that led to the ordinary, governance, and constitutive problems in the first place. This is almost a catch-22 situation where a deficient process must be fixed using a deficient process. The practical problem-solving task is twofold: first, to end old practices that are harmful, and second, to bring about new, more sustainable practices consistent with ecosystem/transboundary management principles.

History of the Committee

The GYCC has a history of more than forty years, which has been described and appraised by various people. David Garber and Jan Lerum, the forest supervisor and head of public relations of Gallatin National Forest, respectively, compiled historical overviews of the committee in 1994.[8] Jack Neckels, Grand Teton National Park superintendent, also described the committee's background at one of its 1996 meetings.[9] The GYCC itself included a historical note in its 2000 *Greater Yellowstone Coordinating Committee: Briefing Guide.* Finally, Mary Maj, GYCC executive coordinator, and Kniffy Hamilton, Bridger-Teton National Forest supervisor, offered a more recent history.[10] Garber, Neckels, and Hamilton all served on the committee. The history below is based on these reports.

1964–1984: Potential

The GYCC originated in 1964 in a memorandum of understanding between the National Park Service of the U.S. Department of the Interior and the Forest Service of the Department of Agriculture. The committee constituted itself probably in response to the 1964 Wilderness

Act and other contextual demands in natural resource management and politics at the time. The committee's originators intentionally limited their arena of action to the Greater Yellowstone Area (GYA), roughly 14 million acres comprising six national forests and two national parks. The memorandum of understanding called for cooperation and coordination in the management of core federal lands in the GYA.

During its first twenty years, the GYCC was minimally active. Meetings were largely informal information exchanges and opportunities to share strategies used in dealing with similar problems. A 1975 meeting, eleven years after the committee was formed, was the first one to focus on issues across greater Yellowstone, according to Jack Neckels. Growing dialogue within the group led to more regular meetings. In the early 1980s, even though committee members recognized that improvements in management policy were needed, they felt that no new legislation was needed to help them or their agencies improve cooperation.[11] They seemed satisfied with the status quo and rate of progress and showed no sense of urgency to bring about any kind of significant change. The major accomplishment before the mid-1980s was development of a comprehensive "Greater Yellowstone Outfitter and Guide Policy" and work on management of grizzly bears, which were listed as a threatened species under the Endangered Species Act in 1975.

1985–1991: Beaten but Unbowed

In 1985–86 the House Interior Subcommittees on Public Lands and on National Parks and Recreation held hearings to examine management policy in the Greater Yellowstone Ecosystem (GYE; larger than the Greater Yellowstone Area by about 6 million acres) and the level of state and federal agencies' coordination.[12] The nongovernmental conservation community, especially the Greater Yellowstone Coalition, helped initiate these hearings. The Congressional Research Service (CRS) analyzed existing data on the ecosystem and agency coordination. Concluding that the agencies showed weak coordination, it recommended that they gather data in a more comprehensive and coordinated fashion, replace the numerous current (sub)committees with a single more comprehensive one, determine the carrying capacity of the GYE, and fix the problem of "lack of coordination" in grizzly bear recovery. As a result, the GYCC's memorandum of understanding was updated in 1986 to reflect these recommendations and call for mutual cooperation and coordination in management of the Greater Yellowstone Area's core federal lands. Congressional hearings in 1986 clearly stimulated the GYCC to action, prompting the committee to undertake a "planning" exercise across federal agency management units. What later became known as the "Aggregation" and "Vision" exercises, two defining events of this period, were agency responses to these hearings. Prior to these two efforts, "each agency played to whatever part of its own constituency [was] in attendance and little was accomplished at each meeting," according to Garber.[13] After 1986 the heads of the agencies changed the committee and its work.

The "Aggregation" exercise commenced in 1985, when the GYCC began compiling data about the GYA through a joint agency planning team.[14] In 1987 the results were published as *The Greater Yellowstone Area: An Aggregation of National Park and National Forest Management Plans,* summarizing current conditions and the extent of resource and management activities.[15] This document showed the incompatibility of goals, information, and activities among the agencies. It also revealed the size and contents of the existing database. Sandra Key, then superintendent of Bridger-Teton National Forest, remarked at a 1996 GYCC meeting that during the "Aggregation" exercise the GYCC had found many conflicts in standards and guidelines across units.[16] She noted too that coordination was needed in fire management, outfitter policy, weed management, and data management. At that time the committee announced its intention to accomplish the needed coordination by overseeing a review and analysis of the issues listed by Congress, those cited in its own report, and any new issues that arose. Overall, little seems to have happened as a consequence.

Second was the "Vision" exercise, a strategic planning activity intended to clarify a vision for the agencies and encourage coordination. Evidence indicates that the agencies decided that this exercise would be the necessary complement to the "Aggregation" exercise and that the two together would be sufficient response to the congressional hearings and CRS report. The agencies' deliberations on these matters are not publicly documented. The exercise resulted in a 1990 draft document, "Vision for the Future: A Framework for Coordination in the Greater Yellowstone Area," which described desired future conditions for greater Yellowstone and laid out management goals for how to achieve this new vision.[17] As described in chapters 2 and 4, this document met with intense public criticism from all sectors.[18] Dave Garber, supervisor of Gallatin National Forest who was then serving on the committee, later observed that the GYCC should have acknowledged that this was truly a planning document and therefore subject to the procedural requirements of the National Environmental Policy Act (NEPA). Garber concluded, "We should have used the NEPA process to

conclude this effort."[19] In hindsight, it is not clear that using NEPA
would have changed the public response or the outcome. Despite what
the GYCC should have done, what it did do was quickly back down
from the ideas put forth in the "Vision" exercise. The widespread public criticism was more than the GYCC could cope with. In so doing, the
GYCC largely abandoned congressional recommendations and its own
goals. It made virtually no credible attempt to defend the "Vision" or
push forward an active program to address criticisms. Committee
members suggested that the "Vision" document was a misguided product created by their staffs before the committee itself had had a chance
to track the process or oversee the final production adequately. Feedback was so negative that four committee members have since publicly said they would never repeat anything like the "Vision" exercise
again.[20]

Since that time the GYCC has put out little effort to promote the
goals it once thought so important. A greatly revised final document,
issued as "A Framework for Coordination of National Parks and
National Forests in the Greater Yellowstone Area" in 1991, removed
much of the specific direction of the first draft and was less controversial. "The GYCC left this era beat-up but wiser," wrote Garber. "As
often happens, the process was much more important than the product."[21] In her historical review, Sandra Key said that the "public got
lost" and that the "Vision" exercise was a "missed opportunity" but an
"honest assessment."[22] Regardless of how this exercise is understood
today, the "Vision" exercise and its aftermath still live actively in the
cultural memory of the committee and the agencies. It is almost universally viewed as a mistake because of the political firestorm it
caused, and the consensus has been to avoid similar exercises at all
costs.

A new memorandum of understanding (discussed below) resulted from deliberations at this time and recreated a group of "unit
managers"—those few people who manage the parks, forests, and
wildlife refuges—to make up the GYCC. Regional foresters and park
directors stayed in the background in their distant offices in Denver,
Colorado, Ogden, Utah, and Missoula, Montana. "The thrashing that
the [unit] managers received during the 'vision' effort changed a group
of managers long suspicious of each other's motives into a close group
that really began to make things happen," according to Garber.[23] With
this streamlined and newly motivated group, the outfitter policy was
revised, and the oil and gas leasing policy was "coordinated," according to Garber's account.

1992–2001: Retreat into Data

During the last decade of the twentieth century, the committee moved from coordination and planning activities to data management. This new focus on intelligence gathering marked a move away from other functions of the management or decision process. Activities related to intelligence or estimation are politically "safe," a common strategy for organizations to undertake when they meet opposition as the GYCC had with the "Aggregation" and "Vision" documents, which had been promotional and prescriptive in nature and had proven costly politically. This change in agency focus and activity, which represented a major strategic shift at the time, has continued to the present.

In 1992 and 1993 the committee sought to develop a common approach to data management, and this activity became the most significant agenda item. During this period the GYCC developed a single elk GIS (geographic information system) layer to test its ability to compile consistent data across all management units and boundaries. It agreed that its member units would not initiate any new inventories unless they were part of a single sharable and integrated database. The committee issued its first newsletter in 1992 and a second in fall 1994, although few have been produced since then. In fall 1993 members of the GYCC agreed to make "ecosystem management" an official goal.[24] GIS data layers were developed, including public land uses, social and economic data, livestock grazing, mineral resources, and others. In 1992–93 the GYCC focused on wildfire management. In late 1993 it focused on weed management policy, another noncontroversial issue.

In 1994 the committee had an agreement in place for information management. Garber's 1994 appraisal noted, "With the benefit of these years of learning and formative efforts the Greater Yellowstone Coordinating Committee is currently making notable progress in joint planning and coordinated management of these public lands. More attainment has occurred in the past three years than the previous twenty. The unit managers team has an unquestionable commitment to implement and help develop the emerging concepts and practices of ecosystem management."[25] That year the committee developed a priority list of issues to address—new GIS data layers that included allocation of public uses (for example, winter visitor uses), analysis and management of social and economic issues (for example, tourism and subdivisions), desired future conditions for the GYA (for example, clarifying objectives), cumulative effects modeling for grizzly bear restoration and rule and regulation consistency, prescribed natural fire, public relations, tactical implementation of coordinated management, livestock grazing, geothermal issues, mineral resources, wildlife disease, air quality, coordinated transportation, and airplane overflights. In early 1994 it targeted wildlands and non-wildland uses.

In 1995 the committee went on a rare winter field trip. Other matters brought to the committee's agenda were addressed in a routine way, as they had been in preceding years. The fall 1996 meeting was an important one. In attendance were the assistant secretary of agriculture for Fish and Wildlife, National Park Service officials from the Intermountain Region, the deputy field director, and three USFS regional foresters. A history of the committee was presented and the group's priorities and funding were reviewed. This meeting revealed how much the GYCC's emphasis had shifted from its earlier goal of developing a common management approach to the present focus on information gathering and GIS technology. At this time, for example, a proposal was presented and discussed for a "national spatial data infrastructure information center and sharing of geographic information systems technology among local, state, and federal governments within the Greater Yellowstone Area."[26] At the same meeting, Jack Neckels presented a summary of GYCC activities and concluded that the committee's history had been "interesting, productive, produced a lot of success, and now GYCC is well positioned for future successes."[27] Finally, the group's consensus at the meeting was that it hoped to be "in a position to make things happen" by early 1997.[28]

Since then the group has held two meetings annually along with a few additional meetings. Most meetings had a session closed to the public, in which the committee looked at its role, activities, and accomplishments. For example, in 1997 the GYCC met in closed session to clarify its goals and identify "issue" areas needing attention (summarized under an examination of goals, below). Since 1998 the committee has continued similar discussions and focused on information, data, and GIS mappings. In 1999 the group worked on winter visitor use and air quality issues and discussed winter elk range protection and migration, large carnivores, water quality, and other issues.

In March 2000 the committee hired its first coordinator, Larry Timchak, a Forest Service official, and defined his job as six tasks. First was serving as the primary liaison for the GYCC board members and adjacent communities, county and state agencies, and other federal agencies that have responsibility for lands in the GYA. Second was to serve as manager of GYCC activities, including analyses and development of support needs (for example, funding, coordination, follow-up of meetings, maintaining contacts). Third was to facilitate development of alternative funding strategies for the GYCC. Fourth was to represent member agencies at the field level, for example, giving presentations to communities, representing the GYCC with national and international figures, media, conservationists, Congress, and key state and local officials. Fifth was to develop and direct implementation of approved priorities and work group initiatives (such as managing and facilitating work of all GYCC-related interagency work groups). And sixth was to coordinate, summarize, and work with others to develop managerial objectives, goals, standards, briefing statements, strategic planning, and other management tools. The coordinator was also expected to monitor and report on progress on all these tasks.[29] Timchak did an excellent job in this multidimensional position, given the context. During this period the U.S. Fish and Wildlife Service added a member to the committee.

In 2000 there was a continuing focus on information gathering and data management. For the first time the GYCC had about $300,000 to spend on "coordinating" activities. Agency officials in Washington, D.C., and regional offices made this "new" money available by rearranging agency budgets. A similar sum has been spent each year since. The projects funded by the committee each year are originated by the member agencies and are carried out largely by agency personnel. In 2000 the committee allocated $310,000 to thirty-one projects, based on proposals submitted by agency personnel. Three criteria were used to distribute the funds: "strategic fit" with GYCC priorities, "application" (benefit to the ecosystem), and how well it might "leverage resources" (matching funds).[30] Examples of projects financed by this increased funding were a Madison Ranger District backcountry weed management project ($10,000), a Yellowstone cutthroat trout interpretative display ($5,000), and an interagency spatial analysis project ($20,000).

In greater Yellowstone there are other federal governmental committees active in research and management, some with a mix of agency and nongovernmental members, which, although mostly or totally independent of the GYCC, also contribute to coordinating management policy. Among these are the Bald Eagle Working Group, Tri-State Trumpeter Swan Working Group, Clean Air Partnership, Jackson Hole Elk Working Group, the Greater Yellowstone Interagency Brucellosis Committee, Greater Yellowstone Weed Group, a Science Group, Clean Cities Group, Winter Use Group, Beartooth Highway Working Group, Whitebark Pine Cooperative Group, a Hydrologist Team, Interagency Grizzly Bear Committee, Northern Yellowstone Wildlife Working Group, and a Fire Management Team. Several groups, committees, and sub-committees do not meet regularly. Even though the GYCC does not have direct authority and control over some of these groups, committees, and subcommittees, membership sometimes overlaps among groups, forming "linkages."[31] Recently, the GYCC has tried to stay in touch with each of them.

In 2001 the GYCC continued its activities as in previous years. It allocated $300,000 to forty-four projects, including winter use monitoring ($4,000), Jackson Hole weed education ($2,500), and Custer National Forest support costs ($7,500). The committee printed a Briefing Guide that described its work and listed current issues and priorities (such as invasive species management, cutthroat trout conservation, fire management, lynx and wolverine surveys, whitebark pine monitoring, and data sharing).[32] This guide also described how the GYCC, as an inter-governmental ecosystem/transboundary management effort, was "transcending boundaries in one of America's most treasured ecosystems,"[33] thus going on record with its overriding goal and intentions.

2002–2006: Regrouping on Goals

In recent years the GYCC has followed a standard pattern in its meetings. Key events have included hiring of a new coordinator, Mary Maj, a well-respected and experienced Forest Service official who replaced Larry Timchak after he became deputy forest supervisor on Custer National Forest. Like her predecessor, she has worked hard at making the GYCC a success, given the context, and has done a stellar job. Her job, like that of the first coordinator, is to be the primary liaison for GYCC members and adjacent communities, county and state agencies, and other federal agencies. She also serves as manager of all GYCC activities, including analyses, development, finding funding, coordination and follow-up of meetings, facilitating interagency issues, and maintaining continual contact with agency support staffs. Overall, she has engaged the members in the committee's activities more than ever before.

At the spring 2002 meeting the committee reconsidered its goals (discussed below). A year later, committee members signed a new memorandum of understanding to "document the intentions and provide a framework for providing public services and responsible land management in a cooperative and coordinated manner."[34] That year the GYCC supported thirty-six projects, including effects of wildfire on grizzly bears' vegetal food in Yellowstone National Park and an annotated bibliography on whitebark pine. In 2003 forty-seven projects were supported, among them a weed mapping project, a vegetation database, and a project using GIS for fire and fuels management. In 2004 thirty-six projects received funding, including a recreation assessment and a project summary report. In 2005 thirty-nine projects were funded, including weed projects, bear-safe food storage, whitebark pine mapping, potable wash station, vegetation mapping. And in 2006 thirty-eight projects received funds, including weed treatment, winter use monitoring, northern goshawk population study, grizzly bear interpretation and education, salt cedar control, aspen restoration, trout inventory, and biennial scientific conference. This funding strategy and the projects selected for support will be examined later. No dramatic departure is expected in the foreseeable future.

Implications

In its long history the GYCC has taken a low profile and operated below the radar of public attention most of the time, with two major exceptions in the 1980s, the "Aggregation" and "Vision" exercises. Its "exclusive executive committee" structure means that it includes only some of the federal agencies and other authorities active in greater Yellowstone. It limits its focus geographically to the land base of its agency members (that is, the national parks, forests, and wildlife refuges), which the committee calls the "Greater Yellowstone Area." Members are the highest-level administrators on their respective federal management units and are highly responsive to elected officials and their bosses, politically appointed or civil service. This influences their decision making. In the mid-1990s, the committee adopted "ecosystem management," at least rhetorically, as both a goal and a means for improving management policy. More recently, the GYCC has represented itself as carrying out transboundary management policy, and it has begun talking about sustainability as a goal. Over its forty-plus years, the GYCC has used a case-by-case approach consistently to give its deliberations and actions substance and form. One example of the GYCC's present focus of attention is the projects that it funds annually.

The Committee's Goals

Goals define the directions that an organization attempts to pursue, according to Richard Daft, an organizational designer and theorist.[35] Organizations exist, in fact, to achieve particular goals, which serve to provide legitimacy, direction for decision making, guidance, and criteria to evaluate performance. More generally, they serve to reduce uncertainty. Goals and goal setting are key features of leaders' capacity, "the ability of actors (individuals, groups, organizations, institutions, countries) to perform specified functions (or pursue specified objectives) effectively, efficiently and sustainably."[36] It is therefore important to look at the GYCC's goals in some detail. Does the GYCC set realistic goals and work practically to achieve them? What are the organization's goals? Official and unofficial goals (that is, the real or operating goals) often differ. If the GYCC's official, substantive goals are to bring about ecosystem/transboundary management, what other unofficial goals appear to be at play? Do the official and unofficial goals and actions all align and support one another to bring ecosystem/ transboundary management policy into existence? Without clear, measurable goals and objectives, it is impossible to determine if progress is being made. Learning about the GYCC's goals tells us more about the situation in which leaders find themselves in greater Yellowstone.

Official Goals

What do leaders say they are trying to accomplish through the committee? Formally advertised, official goals are often rhetorical and promotional rather than statements of real commitment. That is, they are symbolic. The written record contains many references to the GYCC's official goals (see appendix 1). These have evolved over the years, as might be expected, as the group developed experience and a better understanding of the nature of its task and the context in which it operates.

First, official goals are spelled out in various memoranda of understanding and the minutes of meetings, broadly outlining what the group intends to accomplish. This gives a picture of the contours of the problems as the GYCC sees them. For example, the 1985 memorandum of understanding with four signatories (one acting regional director of the National Park Service and three regional foresters) listed fourteen goals:

1. Capitalize on the differences in missions and authorities between the two agencies [Forest Service and Park Service] to achieve in combination a higher level of public service than could be obtained separately.
2. Reinforce existing coordination mechanisms.
3. Encourage cooperation in the field.

4. Consult on potentially conflicting policies, while retaining agency decision authorities.
5. Try to resolve disagreements and refrain from exciting public controversy.
6. Jointly develop research and monitoring procedures.
7. Encourage field units to implement cooperative work jointly.
8. Develop a joint public information and education strategy.
9. Encourage joint public and interagency participation.
10. Coordinate planning and land management strategies within applicable statutes.
11. Conduct evaluations of their own performance.
12. Coordinate schedules to update management plans.
13. Improve compatibility of data bases and information systems.
14. Encourage employee exchanges.[37]

These goals emphasize and reinforce agency and unit autonomy while at the same time calling for increased cooperation and coordination. As a result, they indicate areas of tension and highlight issues requiring trade-offs and compromise to meet all individual agency and shared objectives simultaneously. The history of the committee has been a constant balancing of these autonomy goals versus cooperation goals. In 1986 the GYCC reaffirmed the official goals of the 1985 and the original 1964 memoranda of understanding.[38] This set of stated goals reveals much about the GYCC's operations.

Second, in spring 2003, in a closed meeting, the GYCC again listed goals, "overarching" goals, and "emphasis areas for the next 12–36 months." The three overarching goals were:

1. The GYCC is persistent in its landscape approach to management across physical and political lines, and focuses leadership and resources in support of practices that maintain and enhance the integrity of the Greater Yellowstone Ecosystem (GYE).
2. The GYCC monitors limits of acceptable change in the GYE and uses its collective influence to assure those thresholds are not crossed.
3. The GYCC identifies and prioritizes key areas of agency intersection and develops and implements specific strategies that maximize results for the Greater Yellowstone Area (GYA) ecologically, socially, and economically.[39]

Following on from this, five "emphasis area" goals were produced:

1. Assure the protection of healthy components of the GYE and identify and focus management strategies on the most imperiled components. In the short term, review the previous assessments of ecologically vulnerable components and identify, agree to, and focus management strategies to support them.

2. Develop and deliver consistent messages that are issue specific regarding management strategies supporting the GYE.
3. Promote inclusive, interdependent working relationships with local
and state governments and individual communities. Work to be respected, trustworthy neighbors who are integral parts of their areas.
4. Initiate a "summer use assessment" similar to the completed winter
use product. Use the assessments as a building block to complete Forest Plan revision, travel management, and ultimately, to develop a
collective system for managing the broad spectrum of recreation activities across the GYA.
5. Assertively address the issue of changing land use patterns in critical
areas of the GYA as follows: (a) update and annually produce "the
notebook," (b) facilitate upper-level support in each organization for a
unified GYA land adjustment package, and (c) interact with local government, congressional delegations, and potential partners to explain
the land adjustment package *and* solicit their input and assistance.

We can view all five of these emphasis area goals as getting information (that is, estimation or intelligence) or telling other people about
the GYCC's work (promotional). These goals were reaffirmed and restated in 2005 at the Cody, Wyoming, fall meeting as the GYCC sought
to (1) coordinate planning, strategies, and practices; (2) establish priorities and assign resources; (3) provide a forum for coordination with
others; and (4) minimize duplication of efforts.

Third, importantly, throughout all of the GYCC's work, it has operated within a framework of federal laws controlling it and all its
member agencies, and it has remained committed to complying with
all those laws.[40] The Forest Service is guided by the Multiple-Use
Sustained-Yield Act 1960 (MUSYA), which requires the secretary of
agriculture "to develop and administer the renewable surface resources of the national forests for multiple use and sustained yield of
several products and service obtained there from." The 1976 National
Forest Management Act reaffirmed the MUSYA by mandating that
multiple use is the goal. It also required conservation of viable populations of all desirable and introduced vertebrate species through planning, use of interdisciplinary teams, interagency concurrence, public
involvement, and resource production estimates. The National Park
Service is guided by the National Park Service Act of 1916 (NPSA),
which calls for protecting and using the parks to benefit the American people, with equal emphasis on their enjoyment and protection.
The act grew, in part, out of efforts by the railroads to promote scenic
beauty in order to boost economic tourism, according to Paul Schullery
and Lee Whittlesey, Park Service historians.[41] Other interpretations
of the NPSA have stressed the protection mandate, including the 1963
Leopold Report.[42] According to Richard Sellars, a longtime Park Service

observer, this two-part goal keeps the NPS divided between providing access for public use and protecting our natural heritage.[43] The Fish and Wildlife Service (FWS) is guided by the 1939 Reorganization Act, which created the agency. The FWS is responsible for protection, conservation, and renewal of fish and wildlife and their habitats for this and future generations. No single law guides the FWS. Taken together, these laws and many others form a complex mix of exploitation and protection formulas that federal officials must balance in their daily work.

In sum, this brief account of the GYCC's efforts to clarify its formal, stated goals over the last twenty years shows a clear pattern. The committee feels that goal clarification is an important task and has devoted considerable time struggling to clarify its goals. Numerous meetings have included discussion, reformulation, and further clarification of goals; in some years previous goals were reaffirmed and in other years new goals were formulated and advanced. Overall, the group's goals are mixed in that they support both the autonomy of agencies and units, as well as increased coordination and cooperation among agencies. Goal statements in recent years have emphasized cooperation somewhat more than autonomy, although this does not mean that the autonomy goals have disappeared. The motivation for these repeated goal exercises is perhaps to enhance the legitimacy of the group and clarify its purpose. Perhaps the GYCC and the unit managers who constitute the committee feel the need to make, at a minimum, symbolic gains even though substantive gains are elusive. The fact that leaders are limited to advancing goals symbolically, which is easier and less risky than achieving actual gains on the ground, is indicative of how political the context actually is for these managers.

Unofficial Goals

All individuals, groups, organizations, institutions, and countries have informal, unofficial goals as well as their formal, official goals. Unofficial goals are the real or operating goals, the real story behind the official picture because what a group claims to be doing often differs from what it actually does. Equally legitimate, the two kinds of goals serve different functions, which may reinforce or contradict one another. The discrepancy between official and unofficial goals may come about consciously because a group's members know their unofficial goals won't pass public scrutiny (they are typically never made public and never written down). Or it may come about unconsciously because the members are simply unaware of what they are doing. Consequently, these goals, both conscious and unconscious, must be inferred from an organization's actions. Observing what the group does with its time and resources usually reveals its operating goals. Several aspects of the GYCC's unofficial goal setting are important to understand its behavior.

Official goals may come across as lofty and worthwhile, implying that progress is being made, whereas unofficial goals may point to something else. Sometimes official goals provide cover for unofficial goals. Often in bureaucracies, unofficial goals help to maximize autonomy and discretion and maintain the status quo, current standard operating procedures, and existing leadership patterns, even though official goals may state that the organization is changing, moving forward, cooperating, integrating, and leading in new ways. Unofficial goals function to permit the group to stay in place as a blocking mechanism while at the same time claiming that progress is being made. This is called *dynamic conservatism*.[44] The evidence of the GYCC's agendas, deliberations, and actions suggests that the group operates with these kinds of unofficial bureaucratic goals alongside its official ones. When official and unofficial goals conflict, the unofficial goals, the ones by which the organization actually operates, typically dominate, leading to what organizational designers call *goal inversion* or *displacement*.

Actions stemming from unofficial goals can be contrasted with actions related to official goals, thus highlighting the congruence or incongruence. For example, despite the committee's official goals about cooperation and joint action, one GYCC member said, "The group was not a decision-making body and as a body did not have jurisdiction over anything. Each agency maintains its own mission and policies which we all understand and agree to work within."[45] This statement about unit autonomy appears to be incongruent with the formal, more public goal about cooperation and integration to bring about ecosystem/transboundary management policy. Reconciling official and unofficial goals is one of the hardest tasks for any group, and the GYCC has been no exception.

Comparing official and unofficial goals can reveal the function of goal setting as it relates to advertising (promotion). The GYCC apparently feels that it needs to revisit and promote its formal goals repeatedly to itself, the public, elected officials, and regional and national superiors in the agencies. To be sure, it is a complex task. It is possible that the GYCC constantly redefines its goals because of the complexity of greater Yellowstone, or it may be because the group is unable to tackle more substantive issues. Officially, the GYCC seeks interagency

coordination, data sharing, minimizing conflict and duplication, cooperation on setting priorities, a forum for constructive interaction with
one another, and more—things that almost everyone supports, at least
in principle. Unofficial or real operating goals, however, function to
maintain established patterns of authority and control, unit by unit,
within and among the agencies.

Implications

In order to appreciate the GYCC's goals and how they function, we
need to look at discernible patterns in the GYCC's official and unofficial goals and at how well those goals conform to the challenges the
organization faces. We need to ask whether the GYCC's deliberations
show clear patterns in agenda setting. Who is invited to address the
committee? What is the content and tone of the discussions? And what
conclusions, actions, or follow-ups are actually attended to? It seems
clear that the GYCC is having difficulty clarifying a fully integrated
value position for itself. Its ambivalence over its value commitments
probably reflects the diverse perspectives of the unit managers and the
turbulent context in which the group operates. The GYCC is a pluralistic body and its members, as individuals and as a group, must function in a variety of roles and ways over time.

There are in fact discernible patterns in the goals the GYCC has set
for itself, even though they may be in some ways contradictory or
serve multiple purposes. Chief and perhaps foremost among the goals
is legitimacy. Legitimacy goals, as both official and unofficial goals,
give this leadership group, at a minimum, the appearance of direction,
resolve, and progress that can be publicly promoted. In fact, the record
shows that there has been genuine progress toward coordinated management policy and that this trend may be slowly accelerating. Other
purposes of the group's goals include satisfying people in the external
environment, defining the type of work the GYCC does, helping maintain the GYCC and its members' parent agencies, providing services
and goods, and sustaining individual departments or work units.[46]
These are evident in the GYCC's statements and actions. This goal diversity permits GYCC members to join together and seek improvements in regional management while at the same time continuing on
with business as usual in their home units. It allows them to have it
both ways. On one hand, they can be part of an overall coordinated effort to rationalize regional management policy. On the other hand,
they can continue on with the autonomy they enjoy and operate
within traditional standard operating procedures and bureaucratic

cultures. Regardless of whether goals are directed primarily at substantive outcomes or serve only symbolically, they reveal what values the group embodies or chooses to stand for. At the least, both types of goals are adopted for self-confirming ends, functioning as much as anything else to reinforce existing value outlooks, belief systems (including preferences for how power will be used and justifications for the use of power),[47] and views of the political situation in which the GYCC exists.

Another key question is how well the GYCC's goals and actions address the ordinary, governance, and constitutive challenges laid out in chapter 2. The vast majority of the committee's time, discussion, and decision making concerns ordinary problems on a case-by-case basis. The GYCC and agency staffs spend the bulk of their time in meetings and through their various project-related activities in information gathering, processing, and dissemination activities about ordinary problems. This is an estimation or intelligence activity in the overall management decision process; there is less attention to all the other functions that are also critical parts in this process. The committee also clearly recognizes the importance of governance issues. Its 2003 goals stated that it wants to "promote inclusive, interdependent working relationships with local and state governments and individual communities." Members spend considerable time meeting with elected officials in many venues outside the two annual meetings, and they invite elected officials or their representatives to the meetings. In some of the earlier public meetings, thirty minutes were set aside to interact with the public. It is not clear, however, how governance problems are actually addressed under present arrangements and leadership. Finally, the committee is silent publicly about constitutive problems. There appears to be no direct, explicit discussion about this topic, although there is an implicit appreciation of the constitutive process and its central importance, seen in the committee's goals and actions, whether the group's members are fully aware of this fact or not.

In spite of this apparent lack of fit of the goals to the identified challenges, the goals may yet be successful in a wider sense in bringing about new ecosystem/transboundary management policy. In other words, the existing goals might guide practical problem solving in specific, realistic, and achievable ways. Some of the GYCC's goals and actions promote incremental steps toward new management policy while others maintain the status quo. But none of the GYCC's goals specify in detail improved regionwide management, as compared, for example, with a coordinated ecosystem management effort in the Chesapeake Bay where officials set a goal of reducing pollution by 40 percent within a specified timeframe along with other objectives. The GYCC, for example, said in 1985 that it would develop a strategy of joint public information and education, but it did not specify the content of such programs, establish a program, set measurable objectives, or set criteria by which to measure goal attainment. The committee's broad goals thus do not guide problem solving specifically at an operational level.

One GYCC staff member said, "[the] GYCC doesn't know how much coordination is enough and how much is too little."[48] This observation suggests that not enough attention has been given to goal clarification. The goals are insufficiently clear and consensual to move work ahead on the ground in a way that will bring about ecosystem/transboundary management policy in the foreseeable future. There may be insufficient attention to developing the necessary kinds of knowledge and building a high-quality, deliberative process to arrive at consensual goals. At present, the official goals elevate the rhetoric, while at the same time not making too many promises or too many demands on the time and skills of the members, their staffs, or the parent agencies, nor requiring new resources, restructuring, or increased work loads. They do not call for a significant, sudden departure from current operations. The unofficial goals allow the group to continue operating largely under a "business as usual" formula, support the present unit managers' administration and their "transactional" leadership style, and support the strong interest in agency and unit autonomy and in maintaining leaders' discretion.[49] Although, ideally, incongruence or dissonance among goals should be resolved as much as possible, the complexity of the GYCC's makeup and the situation will not permit that to happen anytime soon.

The GYCC's goals preserve a prominent role for itself and for its constituent agencies and allies in the present dynamic social process that seeks, more or less, an equilibrium of resource flows into society, given society's past and present values. At present the committee meets its many and sometimes conflicting goals (for example, using and conserving resources at the same time) by "satisficing," or accepting present performance minimally "satisfactory" and "sufficient."[50] Such acceptance inhibits more precise goal clarification, which might lead to improved performance, especially concerning governance and constitutive challenges. The GYCC also addresses some problems opportunistically and sequentially, typically on a case-by-case basis. This is a common bureaucratic coping strategy in complex environments and in situations where clear strategic vision is difficult or lacking. The committee would benefit from more precise goal clarification as well as more research on the content of its goals, how they are set, and

what functions they perform in social process as a basis for improvements. Better coordination, including integrated strategies that bring
about a transition toward sustainability, would respond to emerging
values in society and lead to more effective management policy on the
ground.

Integrated Problem Solving

The clear focus of attention in the GYCC's meetings is on ordinary problems, which typically involve difficult or undesirable situations that are
relatively easy to see and understand.[51] Ordinary problems are typically
understood and talked about in conventional language and addressed
within the boundaries of existing bureaucratic thinking, procedures,
and programs. Focusing on ordinary problems piecemeal, however,
does not necessarily bring about an accurate overall definition of the
problems facing greater Yellowstone and its leaders. Creating realistic
problem definitions for each case, as well as aggregating individual
problems into higher-order categories for better strategic oversight and
action, must be part of any successful, coordinated strategy for higher-
order organizational functionality. In the case of the GYCC, this process
of higher-order thinking is critical if large-scale ecosystem conservation
is to be successful. The committee's deliberations, the structure of its
meetings, how they are organized, what topics are discussed, and what
problem-solving tools are used all constitute data about how members
choose the agenda of problems to work on, that is, how they put items
on that agenda and how they orient themselves to these problems once
they are on the agenda. These data provide evidence about how the
GYCC positions itself strategically. Examining the GYCC's present strategy can reveal opportunities to improve problem-solving approaches to
make them more integrative, which is essential in putting the region
clearly on a path toward sustainability.

Meetings

The GYCC'S meetings provide an opportunity for members to interact directly with one another, compare perspectives, and explore
the nature of their shared task. From 1995 through 2004 the GYCC
met twenty-five times or 2.6 times per year over ten years (table 3.2),
although in 1998 it met four times. Each meeting required about two
days. Attendance, an index of members' commitment and priorities,
was consistently high (table 3.3). Of the twenty-five meetings over
this period, virtually every meeting was attended by unit managers

Table 3.2. Dates and Locations of the Greater Yellowstone Coordinating Committee's Public Meetings, 1995–2004

Meetings	Spring	Fall
1995	1/19–20, Bozeman, Mont.	10/5–6, Jackson, Wyo.
	4/26, Idaho Falls, Idaho	
1996	4/3–4, Bozeman, Mont.	10/2–3, Jackson, Wyo.
1997	3/26–27, Idaho Falls, Idaho	10/8–9, Jackson, Wyo.
1998	2/6, Mammoth Hot Springs, Wyo.	7/21–23, Green River
	4/8, Bozeman, Mont.	Lakes Lodge, Pinedale, Wyo.
		10/7–8, Jackson, Wyo.
1999	4/6–8, Idaho Falls, Idaho	
2000	4/5–6, Bozeman, Mont.	12/6–7, Jackson, Wyo.
	8/2–3, Idaho Falls, Idaho	
2001	4/4–5, Bozeman, Mont.	11/15–16, Jackson, Wyo.
2002	5/1–2, Bozeman, Mont.	10/8–9, Jackson, Wyo.
	7/21–22, Trail Creek,	
	Yellowstone National Park	
2003	2/11–12, Bozeman, Mont.	11/4–5, Bozeman, Mont.
	4/2-3, Jackson Hole, WY	
2004	4/6–7, Jackson, Wyo.	11/3–4, Billings, Mont.

Note: Since 2004 meetings have continued at about two per year at various locations in Montana, Idaho, and Wyoming.

(superintendents, supervisors, or managers of the highest rank); only rarely did they send lower-level staff as their representatives. Clearly, members feel the committee is important to them. Non-GYCC members and people reporting to the GYCC at each meeting varied from four to thirty or more, depending on location, topics, weather, and other factors.

The structure of a typical meeting is flexible but follows a general pattern. It begins on the first afternoon with a public session running from about 1 P.M. until 5 P.M. On the evening of the first day, committee members typically go to dinner as a group by themselves. The meeting resumes the next morning from 8 A.M. and runs until 12 noon in a session closed to public. Typically, eight to ten topics are addressed in the public session on the first day, with fifteen to forty-five minutes allotted to each agenda item. This time is largely taken up by presenters reporting on their assigned topics, leaving little time for in-depth questions or discussion from members or anyone in the audience who may be recognized by the committee and granted permission to speak, although issues sometimes stimulate brief discussion among committee members. The public is not invited to participate directly except at the end of the day during the thirty minutes (or less) set aside for that

Table 3.3. Pattern of Attendance at Meetings by Greater Yellowstone Coordinating Committee Members

Management Units Represented	1995 Jan., Apr., Oct.	1996 Apr., Oct.	1997 May, Oct.	1998 Feb., Apr., July, Oct.	1999 Apr.	2000 Apr., Aug., Dec.	2001 Apr., Nov.	2002 May, July, Oct.	2003 Feb., Apr., Nov.	2004 Apr., Nov.	Score/Maximum Possible
Shoshone National Forest	U, 3, 3	2, 3	3, 3	3, 3, 3, 3	3	3, 0, 2	3, 3	3, 3, 2	3, 3, 3	3, 3	66/75
Gallatin National Forest	3, 3, 3	3, 3	3, 3	3, 3, 0, 3	3	3, 3, 3	3, 3	3, 3, 3	3, 3, 3	3, 3	72/75
Custer National Forest	2, U, 2	2, 2	3, 2	3, 3, 3, 2	2	2, 0, 3	3, 3	3, 3, 3	3, 3, 2	3, 3	60/75
Beaverhead-Deerlodge National Forest	2, U, 3	3, 3	3, 3	2, 3, 2, 3	3	3, 0, 3	3, 3	3, 0, 0	3, 3, 3	3, 3	60/75
Bridger-Teton National Forest	U, 2, 3	2, 3	3, 2	3, 3, 3, 3	3	3, 0, 3	3, 3	3, 3, 3	3, 3, 3	3, 3	66/75
Caribou-Targhee National Forest	3, 3, 3	3, 3	3, 3	3, 3, 3, 3	3	3, 3, 3	3, 3	3, 3, 3	3, 3, 3	3, 3	75/75
Yellowstone National Park	2, 3, 3	3, 3	3, 3	3, 3, 3, 3	3	2, 3, 3	3, 3	3, 3, 3	3, 3, 2	3, 3	72/75
Grand Teton National Park	3, 2, 2	3, 2	3; 3	2, U, 2, 2	2	3, 2, 2	3, 3	3, 3, 3	3, 3, 2	3, 3	62/75
Fish and Wildlife Service					3	3, 3, 3	3, 3	3, 3, 3	3, 3, 3	3, 3	42/42

Note: Attendance was scored as follows: U = incomplete record, 3 = superintendent/supervisor/manager, 2 = subsuperintendent/subsupervisor/submanager, 1 = technical substitute = 1, 0 = no representative attended.

purpose, if time is allotted on the agenda for public interactions at all. The meetings are presided over by a chairperson from among its members, a position that rotates every year. Each member has one vote. Sometimes the meetings are announced in the local and regional newspapers a week in advance. In earlier years, the format was a single day-long meeting, and all these topics were compressed into that shorter period.

The winter 2000 meeting was typical. The GYCC met on December 6 and 7 in the Jackson Meeting Room at the Wort Hotel in Jackson, Wyoming (table 3.4). This meeting was preceded by a half-day, closed executive session in which three decisions were made. The public meeting began at 1 P.M. with a welcome and introductions by the chairperson. The decisions were announced. First, the members supported completion of a briefing guide and creation of a Web page for public relations purposes. Second, they decided to support compilation of information from forest, park, and refuge management units in the GYA. Maps were to be created using GIS analysis. They committed $25,000 in support of the technical expertise on GIS and data management, storage of information, and creation of graphics and necessary presentations. Third, the members allocated $30,000 to collect data on recreational trends and conflicts and map results in GIS annually. The meeting was summarized in seven pages of minutes, about ten lines per agenda item.

The minutes are the only public record of what is accomplished at the meetings, whether all members participate in discussions and decisions, whether some members dominate, and what roles members played (critic, innovator, observer, or others). Minutes, taken by an agency staff member or the coordinator, vary in content and length from meeting to meeting, so they limit textual analysis. They summarize rather than detail the presentations but do occasionally quote members' comments. All in all, they give a good picture of the overall discussion on a topic to the extent one took place. Points of disagreement and debate, which are rare, are seldom mentioned. Also, the minutes are not comprehensive and so do not include reports on all presentations and discussions, nor do they summarize all discussion or attribute statements. They do not identify which members spoke, for how long, on what topics. As a result, it is not possible to quantify data across the reports to characterize these attributes.

Attending the open sessions as an observer and recording these types of data do provide a substitute, unofficial record. At the December 2000 meeting, which I attended, there was no conflict in any of the discussions, no one member dominated discussions, and few members

Table 3.4. Agenda for Greater Yellowstone Coordinating Committee meeting, December 6–7, 2000, in Jackson, Wyoming

Wednesday, December 6

1:00	Welcome and introductions	Jerry Reese, chair
1:15	FY 2000 project review and FY 2001 project solicitation	Larry Timchak
2:00	Yellowstone cutthroat conservation strategy	Bruce May
2:30	Jackson elk-bison management plan update	Barry Reiswig
3:00	Break	
3:30	Whitebark pine monitoring project	Bob Crabtree, Yellowstone Ecosystem Studies
4:00	Invasive species management	Fred Lamming Larry Timchak
4:30	Public question/answer period	
5:00	GYCC managers reception, Gold Piece Room, Wort Hotel, cash bar	

Thursday, December 7

8:00	Inland West Water Initiative strategy Air partnership update	Greg Bevenger Mark Story
8:45	Fuels/fire strategy for GYA	Len Dems
9:15	ESA petition updates and possible management implications trumpeter swans wolverines	Terry McEneaney Jim Claar
10:00	Break	
10:30	Data management recommendations	Henry Shovic
11:00	Managers' roundtable	All managers
12:00	Adjourn	

asked questions of the presenters or offered their own perspectives on topics. The meeting was largely an informational one, with speakers making presentations using slides, overheads, or PowerPoint projections. Most of the information presented was about historic trends in the subject under discussion. About ten members of the public attended off and on over the two-day meeting. The public question/answer period brought out six questions, usually on matters of fact that needed clarification. There was neither substantive give-and-take interaction nor in-depth discussion between GYCC members and the public. Overall, the meeting showed little analysis of trends or exploration of explanations, future projections, problem definitions, or alternatives to solve implied or perceived problems. Many meetings that I attended showed a similar profile. We must ask whether this kind of meeting or interaction brings out the kind of high-quality, integrated problem solving needed to set or meet official goals of ecosystem/ transboundary management. It is possible that all integrated problem-solving deliberations occurred in the closed sessions, but my discussions with attending GYCC members and the minutes from those meetings show that this was not the case.

Agendas

The GYCC focuses its attention on eight to ten cases at each meeting. Selection of agenda topics seems to be based on a combination of pressing, acute issues (for example, a petition to list Yellowstone cutthroat trout under the Endangered Species Act) and longer-term, chronic problems (including recreation uses and perhaps lack of appropriate oversight and management). Very few of the many problems discussed in chapter 2, particularly governance or constitutive problems, are brought to the meetings as formal agenda items or otherwise discussed publicly. Nevertheless, they are always present below the surface and in the far background. No agenda item, regardless of import, scope, or immediacy, receives more than thirty to forty minutes of attention. Again, the December 2000 meeting was typical. Among the topics was elk and bison management. Half an hour was devoted to an update on elk and bison management in Jackson Hole and the upcoming environmental impact statement under the National Environmental Policy Act. The risk of catastrophic disease and loss of the herd under the present artificial feeding program in Jackson Hole was emphasized. The environmental impact statement process had begun, although the roles of state and federal agencies remained uncertain. The meeting was adjourned at 5 P.M. and followed by a reception open to the public. Receptions were traditional during those years but have since been discontinued. On the second day, the agenda turned to other topics—mapping of watershed conditions on national forests and in Yellowstone National Park and similar topics.

Fifty-seven agenda topics were recorded from 1995 through 2004 (see table 3.5 and also appendix 2). From six to twenty topics were addressed at each meeting, but typically about eight to ten were covered. About half the management problems on the agenda at each meeting were new and had not been previously discussed. There is no separation of old from new business. This reflects the shifting focus of attention of members as well as the rapid turnover or cycling rate of emerging or persistent problems in greater Yellowstone. Some subjects (including winter use assessment, large carnivore management,

Table 3.5. Summary of Greater Yellowstone Coordinating Committee Agenda
Items, 1995–2004

Discussed at:	Agenda Items
7 meetings	Direction, philosophy, expectations, structure, procedures for GYCC Winter use management Fire management
6 meetings	Grizzly bear model/study New chair GYCC priorities Waterways/watershed/fish (cutthroat trout)
5 meetings	Funding Project review/solicitation
4 meetings	Air quality and transportation Recreation use/fees Land patterns Whitebark pine
3 meetings	Science working group Data collection/management Wilderness/wildlands
2 meetings	Information management Partnership opportunities Heritage Trust Weed management Fund raising Soils mapping Invasive species Jackson elk and bison management Wildlife
1 meeting	Social assessments Budget Vegetation mapping/data base/training NPS fee collection Rocky Mountain Elk Foundation Open space Public information Land acquisition Campground reservation system Forest orders about food storage Diversity in the workforce Commercial use permitting Sheep management Backcountry management Community monitoring program DNA microbial research Systems engineering Cellulose conversion

Table 3.5. (*continued*)

Discussed at:	Agenda Items
	Interdisciplinary problem solving
	Clean Cities Coalition
	GYCC records/archives
	Trumpeter swans
	Wolverines
	Yellowstone Teton region system
	Strategic research and science strategy
	Memorandum of understanding/charter/amendment
	Interagency cooperation on federal highway projects
	Inventory and monitoring
	Human safety/sanitation
	Forest land management plan
	Private conservation
	Forest Service centennial celebration

Note: Information compiled from agenda topics listed in meeting minutes, 1995–2004. Data missing from January 19–20, 1995; April 25, 1995; April 3–4, 1996; and May 14, 1997.

and data collection and management) were talked about year after year, although most topics were discussed only once. Overall, the meetings, agendas, presentations, and discussions reveal how the GYCC orients to problems.

Problem Orientation

How people orient to problems determines whether the problems can be understood in the first place, much less solved. Being fully "problem oriented" maximizes chances of success as opposed to being "problem blind" or "solution oriented," both of which invite failure.[52] Some strategies for understanding problems and finding solutions are better than others.[53] Approaches have been developed that are much more contextually sensitive to real-world concerns.[54] Such approaches encourage both a look outward at "problems" and a look inward at "problem solvers, critical thinking skills, and problem solving assumptions and strategies."[55] The GYCC's problem-solving behavior can be compared against these accepted best practice standards. Several conclusions present themselves from the comparison.

Although the GYCC appreciates that the greater Yellowstone ecosystem is interconnected in many ways, when faced with actual problems it typically "decouples connections" and focuses on ordinary cases as they occur in individual units. Whether involved in GYCC deliberations or in agency operations, members tend to look at such problems one at a time. This is a common coping strategy when faced with a complex context with many problems of diverse kinds. Most of the complexity is "defined, ignored, or managed away" using this strategy to make the analytic and communication tasks more tractable, given the bureaucratic boundaries that are ever present.

Problems are seldom fully or practically defined in a way that would provide a basis for inventing potential alternatives for consideration and decision. The five tasks or operations of problem orientation, as discussed in chapter 2, are not carried out completely or systematically for any one ordinary problem, much less for any higher-order categories of ordinary problems or any of the governance and constitutive challenges. As a result, weak problem solving dominates and leads to an array of fragmentary solutions that give a false sense of accomplishment.

Because little attention is focused on understanding problems in a fully problem-oriented and comprehensive way, larger patterns go undetected and the true nature and extent of problems go unappreciated. There are many important issues besides the ones that have made it onto the agendas, yet only certain specific cases have received the committee's attention, however briefly. The GYCC's strategy precludes more comprehensive regional assessment and exploration of ways to meet the more systematic problems The real risk of taking an overly selective, case-by-case approach to ecosystemwide problems is that it falls short of providing the needed strategic overview and insight. Of course, taking only a systemwide, comprehensive overview across all challenges could be analytically overwhelming, unless the GYCC is willing to commit the time and resources to pursue this kind of an approach. Without an adequate problem-oriented approach that balances selectivity with comprehensiveness, however, a realistic picture of management policy in greater Yellowstone is impossible.

Implications

Greater Yellowstone is an arena or zone of conflict that arises from the interaction of population, environment, and resources. Many people compete for available resources and this leads to conflict in many cases. In this arena and context, the GYCC has chosen a stripped-down approach to the setting, the conflict that it contains, and its own role. Several additional conclusions seem warranted about the committee's strategy for problem solving.

First the committee is directed too much inwardly, with meetings devoted largely to presentations to the committee from staffers and rarely with efforts to reach out to responsible NGOs or academics for fresh perspectives and new insights. These staff reports mostly concern status of knowledge, ongoing activities (for example, fire data from previous years, status of soil maps, remote sensing technology), or in a few cases the implications of various management initiatives. But there is virtually no analytical, problem-oriented discussion among members, with presenters or the audience, and few actual conclusions are reached publicly. The case reports are not evaluated formally against the GYCC's official or unofficial goals and reported on publicly. This approach keeps the contextual and political content of the cases in the background, unexamined, while the focus remains on technical matters and targets of possible management intervention or manipulation. As a result, governance and constitutive challenges are addressed only implicitly and even then incompletely. Few records exist for closed executive sessions, where more substantial deliberations may take place. It would be helpful if these were made public.

Overall, the committee's approach emphasizes information sharing about trends (historical data and impressions), but it significantly neglects the other, equally important tasks of clarifying goals, analyzing conditions and causes of the observed trends, projecting trends into the future, creating alternative problem definitions, and fully exploring alternatives or options based on their findings. The net affect is that the committee emphasizes substantive rationality (that is, technical matters, the facts of the matter at hand, or at least expert reports) and minimizes procedural rationality (such as ensuring a thorough, problem-oriented approach by explicitly attending to the essential tasks of problem solving). This gives the impression to GYCC members, agency staffers, and the public that the GYCC is addressing the cases at hand effectively.

The GYCC's case-by-case approach is a conventional, safe strategy, widely used in bureaucracies. It serves several functions.[56] A case-by-case approach focuses unit managers' attention on the ecosystem "out there" and its biophysical elements (thus minimizing human process and contextual issues). This technical construction sets problems up as targets of manipulation and deflects attention away from the human process, the contextual issues involved, or the GYCC's own problem-solving approach. It allows managers to be somewhat flexible and allows them some wiggle room in overall goals and management actions. Managers don't need to get the answer completely right as long as the chosen solution more or less addresses the specific problem. Goals and actions are interconnected, complex, dynamic, and socially constructed, so maneuvering room is often needed to address problems. This kind of flexibility can be good when first addressing a category of problems, but if it persists, it can lead to sloppy thinking and failure to arrive at an overall consistent conclusion. Each case is treated on its own merits, (1) in terms of whether management achieves its stated objectives, (2) in terms of the implementation record of like management efforts, and (3) in terms of where monetary savings could be realized if the management were undertaken most cost efficiently. This assumes that someone does overall monitoring and assessment, which is likely not the case. Overall, this gives the GYCC flexibility, autonomy, and discretion. This approach draws more from the informal experience of managers than from the more formal models of, for example, ecology or policy process. The approach that the GYCC currently takes is clearly not a form of active, adaptive learning in any formal sense. Over time, regardless of outcome, cases can be portrayed in favorable terms. A case-by-case approach, however, makes it hard to generalize lessons for systemwide improvements unless an active learning system is in place to harvest lessons, disseminate them, and apply them widely.

The GYCC does not meet often enough or employ a dedicated staff large enough to achieve its formal goals. If the GYCC assumes that the challenge is nothing more than a series of ordinary issues, which seems to be the case, then a relatively simple extension of present management arrangements could be deemed successful. In that situation, success would be achieved incrementally through agency bureaucracies, so relatively few meetings each year would probably suffice. If, however, the challenge is viewed as a more complex policy task made up of interactive ordinary, governance, and constitutive challenges, which the GYCC seems not to assume, then new goals, organizational structures, and knowledge and skill sets, as well as active oversight, feedback, and active learning are needed. In this latter scenario, more frequent meetings and a larger dedicated staff would be essential.

At policy levels, issues are considered highly coupled if they are really connected problems with governance and constitutive implications. Ordinary problems are usually surface manifestations of some deeper governance or constitutive issues. Yet the GYCC emphasizes a "decoupled" approach. The GYCC's focus on ordinary problems, case-by-case, accords little attention to linked governance and constitutive problems. This means that coordinating management policy at high systems levels cannot be successful under the present strategy. Currently, management is carried out in the simplest way; for example, all the units coordinate by uniformly conducting a weed survey/control effort. Such actions are not a substitute for the needed higher-order policy and management.

Overall, the GYCC's problem-solving strategy gives attention to selected biophysical trends; partial explanations of trends; and single, simple solutions that are selected, rejected, or deferred. This underattends to the actual problems at hand, which are mixes of ordinary, governance, and constitutive problems. These strongly interconnected problems cannot be addressed with an emphasis on case-by-case, ad hoc approaches. If leaders miscast what is involved in a problem, then they will likely misallocate resources and invite failure. They may choose a course of action because it is the most appealing or expedient at the time,[57] or because it casts them in the best possible light, produces the fewest conflicts, or otherwise permits them to view themselves and their work favorably. The GYCC's approach of decoupling may achieve short-term reductions in environmental uncertainty and increase technical and political manageability of the problem-solving process. But the decoupling strategy reinforces pressure to stay decoupled for the sake of maintaining this manageability. Furthermore, it undermines the need to consider all couplings, optimize, and make trade-offs systematically and consciously across the entire policy and management system.

Conclusions

Several conclusions and preliminary recommendations seem evident from analysis of the committee's history, organization, goals, meetings, deliberations, and problem-solving strategy. These help diagnose the present situation and suggest what might be done to help leaders and management policy in the future. First, the GYCC's forty-plus-year history has shown a very slow trend in the direction of increased integration and coordination in managing greater Yellowstone. The diversity of management policy issues, the accelerating rate at which these rise to the surface, and their significance all require a more collective, integrated response. The job of leaders is to find appropriate responses to public problems. Instead of attending to leadership systematically, the GYCC's typical response to problems is to ask for more information or study (such as a GIS map of fish habitat), improved data management, increased funding or resources, or more support for "safe," noncontroversial projects (such as weed eradication). As a result, the higher-order, problem-solving tasks that we might expect from such a high-level leadership group, as well as the deep policy reflection on which its operations should be based, are nowhere evident in the GYCC's public deliberations or actions.

Second, the GYCC's goals are insufficiently clear or consensual. The members give insufficient attention to developing or using the kinds of knowledge needed to build the high-quality, deliberative process necessary to develop goals and procedural and substantial progress on the ground. The group could elevate its efforts to harmonize conflicting goals and reconcile traditional resource uses with demands for more sustainable management policy. Presently, the GYCC is operating under two missions, an unofficial one to maintain the status quo and an official one to transition toward sustainability. Unfortunately, the second of these missions is receiving short shrift as the present group has inherited a bureaucratic system and a set of resource use rules that are highly problematic over the long term. With smart, principled leadership these missions can be reconciled and made to serve management policy better for the common good. To achieve this potential, the GYCC can shift its focus of attention and get a larger, more highly skilled staff dedicated to help them transition toward sustainability.

Third, the GYCC is struggling to orient realistically to the multidimensional understanding and addressing of problems. The challenges cannot be understood in a realistic contextual way, much less successfully addressed, with incomplete, conventional problem-solving concepts and tools. Solving problems by examining how they have grown or shrunk over time, determining causes of the problems, ascertaining their probable future status, deciding whether a significant problem exists given explicit management goals, defining problems reasonably and practically, and inventing alternatives and evaluating them to address problems as defined is a proven strategy for solving problems.[58] Selecting the option judged to be the best, implementing it, and monitoring its impact are follow-up steps in any complete learning approach to problem solving. These activities, when carried out empirically, consciously, creatively, and interactively, allow problem solvers to orient to problems in reasonable, practical, and moral ways. As policy researchers Garry Brewer and Peter deLeon have noted, other "approaches may appear to offer simpler or easier solutions, but each usually turns up lacking in important ways—not the least of these being their relative inability to help one think and understand, and hence to become a more humane, creative, and effective problem solver."[59]

4 Leaders—Cooperation and Demonstrations

In addition to problem solving, leadership requires cooperation, which includes joint decision making, resource sharing, personal relations, communication, and reciprocity.[1] By definition and by necessity, cooperation is at the heart of any successful ecosystem/transboundary effort. It is certainly required for transitioning toward sustainability, a task that is large, complex, and dependent on joint action. Cooperation on this task requires both basic and high-level coordination and integration. Finally, leaders must also demonstrate to people what they want to achieve across the region by creating actual, on-the-ground, working examples of the processes, structures, and values they hope to cultivate.[2] It is vital that they show, through actual examples, the value-added benefits of new management policies directed toward sustainability. Creating "demonstrations" (that is, prototypes and best practices) at small and large scales is one widely recognized way to bring about constructive change. This is the best way to show confidence in partners and thus encourage cooperation.

In this chapter I examine cooperation and demonstrations by the GYCC. After beginning by looking at how the committee operates among it own members, within its own meetings, and with the public, I go on to describe how the committee participates in decision-making processes, what standards it uses, and how it attends to common, foreseeable, decision-making pitfalls. Next, turning to demonstrations, I examine whether the committee itself amply constitutes a demonstration of the leadership and new policies that are needed for a transition to sustainability. By looking at the specific demonstrations the GYCC chooses to sponsor, its accomplishments and milestones, how it spends its money and uses its staff and subcommittees, the implications of the committee's cooperative behavior and demonstrations are considered.

Cooperation—Measures and Patterns

Leaders must build cooperative working relationships that maximize the odds for success. Given the GYCC's official goals, this means that members must cooperate with one another, with professionals inside and outside the agencies, and with the public to accomplish what they cannot achieve separately. Successful cooperation must be sensitive to both internal and external requirements, that is, to the people and situations within and outside the community. Cooperation can take a low-level form, such as sharing equipment, or it can take a high-level form, such as integrated, interunit planning, implementation, and evaluation. All levels are possible and necessary. Leaders should keep the big picture in mind while targeting specific opportunities for progress in their cooperative efforts. Without high-level problem solving and cooperation, attention to low- and even mid-order tasks will not add up to successful regionwide ecosystem/transboundary management policy. These various possible forms of cooperation raise the question of how and with whom the GYCC does and should cooperate.

Cooperation within the Committee

One index of the committee's posture on cooperation is which groups it hears from and deliberates with at its public meetings. Direct observations of meetings provide a window into intermember, interunit, and interagency cooperation as well as cooperation beyond the committee and the agencies. The tenor of meetings that I have attended has been congenial, and members interacted with one another in a respectful way. Laughter and friendly relations were common. Meetings flowed smoothly, often informally, and the attitude of members was positive. On the surface, there was cooperation via social interactions, data and information exchanges, and through the executive secretary position.

Borrowing a six-level system to measure cooperation among nations, developed by the European Union (EU), can give us more insight into the committee. This system recognizes six levels: (1) no cooperation; (2) some information sharing; (3) notification of actions, emergencies; (4) frequent communication and meeting, active cooperation on multiple activities; (5) regular meetings, coordinated actions; and

(6) fully integrated, planning, cooperation on management, and joint decision-making committees. My application of this index to the GYCC shows that its cooperation has been a mix of levels 2, 3, 4, and 5, depending on the issue and time. These forms of cooperation are valuable because they build trust and friendships that allow members to learn from one another and to pick up the phone between meetings and talk to one another. Overall, GYCC meetings show a low- to mid-level form of cooperation, which, if used well, can be a base to create higher forms of cooperation over time.

Researcher Dorothy Zbicz adapted the EU's system in a study of transboundary management worldwide.[3] She noted that many transboundary programs get stuck in one of the lower levels because they institutionalize interactions, and once that occurs, it is almost impossible to get the members to examine whether the form of cooperation in place actually meets their goals.[4] Many times groups fail to advance cooperation even when it is possible to do so because advancing cooperation requires active, innovative leadership.

Cooperation at Meetings

Another index of the GYCC's cooperation with others comes from reviewing the agendas and minutes of all the public meetings (and supplementing them with direct observations when possible). This tells us whom the GYCC hears from, their affiliations, and the subjects presented. Typically, twenty to thirty people attend the meetings, including committee members, about three-quarters or more of whom are agency employees. Staff members of U.S. senators and congresspeople also frequently attend. Attendees are mostly employees of GYCC member agencies, employees of other federal agencies (such as the U.S. Geological Survey), contractors with the federal agencies, or people who have an established working relationship with the agencies (such as The Nature Conservancy), although it may be informal. Among the attendees who have spoken to the committee in recent years were: Dr. Robert Crabtree, a scientist with the Yellowstone Ecosystems Studies, who spoke on hyper spectral imaging of whitebark pine; Dick Jachowski, director of the Northern Rocky Mountain Science Center (USGS), who spoke on the Greater Yellowstone Science Agenda; and Ron and Alex Diekmann, project managers for the nonprofit Trust for Public Lands, who spoke on land pattern strategy. Additionally, SuzAnn Miller, contractor on the Changing Values, Changing Times project, a private project, spoke on assessment of demographic and economic changes in Montana relative to expectations about natural resource management; Laura Hubbard, the Nature Conservancy's Greater Yellowstone program leader, spoke on private land conservation in the GYA and policies being pursued by the Nature Conservancy; and Jan Brown, executive director of the Yellowstone Business Partnership, gave an update of how the partnership was developing and their future plans. Dennis Glick, a regional representative of the Sonoran Institute, summarized his group's program. There have been others.

Comparisons of recent meetings (2005 and 2006, for example) to meetings before 2000 suggest the same kind and frequency of contact with individuals and groups. The basis for receiving an invitation to speak before the GYCC, however, is not clear, and it would be helpful for the organization to spell out how nongovernmental presenters are chosen as well as how report topics are chosen. The potential is high for greater future cooperation within the GYCC and with people currently outside of the GYCC's sphere of activity.

Most attendees presenting or otherwise speaking at GYCC meetings are government employees. There is roughly a ten-to-one ratio of presenters who are part of the GYCC, government agencies, or those with a close working relationship to the agencies versus non-GYCC/nonagency presenters. Clearly, meetings are dominated by the GYCC and its agencies' personnel, which reflects the executive committee's exclusive makeup, focus of attention, and operations. In short, the committee members mostly listen to reports from their own employees and contractees. This is because the GYCC has chartered subcommittees and working groups who are asked to report back to GYCC on their progress. Because the GYCC meets only twice a year and the agendas are full, there is little time to do more.

Cooperation with the Public

Yet another measure of the GYCC's cooperative behavior is the time the group allots to interact with the public in meetings. The committee has said repeatedly that one of its goals is to interact with the public through its meetings. Among the stated goals are to provide a "forum for interaction with other federal, state, and local agencies and private organizations and the public" (goal 5, April 1996) and to "interact with local government, congressional delegations, and potential partners . . . and solicit their input and assistance"(spring 2002). Other goals are to inform and educate the public. Such statements set up the expectation in the public that there will be meaningful public participation, perhaps even deliberative engagement, and possible cooperative problem solving.

Upcoming meetings are sometimes announced in local newspapers a week in advance, and some sessions are open to the public. Although for a few years, especially in the late 1990s, the committee allowed thirty minutes at the end of the open session for public input, many meetings left no time for formal public interaction for whatever reasons, and most recently the agendas themselves have not scheduled time for public discussion. People who attend sometimes speak up throughout the deliberations, but this is rare. Minutes from the October 8–9, 1997, meeting are typical:

> County Planning. Again, not too much discussion on this item. The GYCC would like to coordinate our federal planning efforts with those of the states, counties, and local municipalities. The group then discussed public involvement with the audience. Debbie [Austin, a GYCC member] asked the public how the[y] would like to be involved. Following are some of the responses. We would like regularly scheduled, advertised meetings. If you have nonpublic meetings, don't appear that you are trying to hide something from us. We must begin with education of the general public. . . . Tim Clark stated that GYCC should be involved in learning how to be involved with public decision making to secure common interest. If we have a big issue (e.g., the Winter Use Assessment), we need to make sure that the public [i]s informed throughout the process. We can put information in interest group newsletters. Can link through the Internet to the gateway community Web sites. Steve Primm [an independent policy researcher] stated that he has information on fire in the Greater Yellowstone.
>
> Concluding Statements. We asked the public again how they would best like to be involved with the GYCC in the future. Those present stated that they would like a summary of the meeting notes and felt that the social hour was very worth while.

To date, most of these public expectations and requests for cooperation have not been met on any regular basis. It is challenging to find ways to have substantive discussions with the public in an open government format while at the same time encouraging candid, creative-thinking, and risk taking by participating leaders, who are understandably concerned about making a misstep in a public setting.

The trend is for the GYCC to have meetings, or at least the second morning of the meetings, as "executive sessions," to which neither the agency presenters nor the public is invited. This may help with interpersonal interactions among GYCC members, but in the context of high-level, public leadership, it does little to promote the GYCC's agenda with the public.

Implications

If the committee is serious about meeting its own goals, a spirit of cooperation should be evident in all the GYCC's operations—at its meetings, in frequent interaction between meetings among subgroups and the GYCC, and in its interactions with the public. Ideally, we should find at meetings abundant evidence of cooperative efforts, persuasion, and mutual assistance at all levels and among diverse parties. Likewise, we should see that the force of a better argument, based on evidence, best practices, and lessons from public policy, prevails over the use of power, coercion, or evasion. Disagreements should be more or less resolved using the appropriate level of cooperation and expending the minimum of social capital.[5]

What patterns do we actually find in the committee's activities? First, cooperation in greater Yellowstone seems to be based on two mutually reinforcing motives. One is a genuine desire by the members to manage the units and the region well, suggesting that management should take place via mutual adjustment wherein one part acts and the second adjusts accordingly. A second motive for cooperation seems to be the need for a unified defensive policy. A united front among the cooperating agencies helps them resist political and legal challenges and helps all members stave off challenges to their authority and management. It gives unit managers and members of the GYCC a good rationale for their decisions. In this mix, there is also a drift to the lowest common denominator: the optimal, most innovative, long-term solution is often sacrificed to the need to reach a position acceptable to all participants. For example, the major, high-order tasks of regionwide leadership and problem solving are simply not explicitly addressed. Cooperation seems sometimes to take the form of uniformity and conformity rather than individualized mutual adjustment. Rather than establishing cooperation as a strong, high priority action goal, cooperation in the GYCC, to the extent that it occurs, seems to be motivated more by mutual convenience and by the need for increased legitimacy as a general matter. This drift toward safe solutions becomes even more pronounced as public trust in government, and hence social capital, decline in the United States. As a consequence, the GYCC's cooperative efforts tend to be conservative and function to bring about management gains at the margins. For example, there is no discussion of significant departures from business as usual. Cooperation is largely limited to established interagency and interunit relations. Reliance on this pattern of cooperation reinforces the present outlook and maintains current agency interrelationships. It has resulted in better management than if the agencies had been working on their own, but too little creativity is brought to bear at GYCC meetings given the ordinary, governance, and constitutive problems that the committee faces. As a result, more innovative opportunities for cooperation are overlooked or receive limited attention. This is one of many forms of self-blocked learning; the most difficult of these to overcome are the "blinders" caused by a group's operating assumptions, the way it structures its operations, or various political reasons. To a degree this reflects the temperaments and variations among GYCC members, some of whom have been creative, innovative, and forward thinking whereas others have not been.

Second, there is a clear pattern of framing all problems as scientific or technical rather than sociological or political. The meetings provide a window into how the GYCC and the agencies go about gathering, processing, and disseminating information. The meetings focus on task-related information, largely limited to historical trend data and thus neither fully problem oriented nor contextual.[6] Because the focus is on case-by-case management, often in a fragmented manner and often in response to threats (political or otherwise), and because time for analysis and discussion is limited, it is not surprising that presenters are highly selective in what they present. Typically, presenters are scientists, technicians, or managers, thus giving the impression that the GYCC's deliberations and decisions are scientific and technical, at least in an ecological sense. With a few exceptions, the social and integrative sciences are never mentioned at meetings, nor is there ever a call or discussion to bring them into the deliberations. The exceptions include the invasive species and recreation groups, which seem to be more attuned to social context. Again, this limits the range of information heard and circumscribes the discussions—another form of self-blocked learning.

Third, the GYCC has developed neither the patterns of cooperation effective for creating and protecting social capital nor the strength and density of relationships needed in a social network, especially beyond the GYCC itself. Human communities with dense webs of relationships and strong, trusting connections possess high social capital. The advantages are obvious—social capital, "the collective value of all 'social networks' and the inclinations that arise from these networks to do things for each other," according to Robert Putnam, author of *Bowling Alone,* allows a community to be more productive.[7] Unfortunately, there is little active or specialized effort within the GYCC to increase cooperation with the public. An exception is the winter use assessment that included some public outreach. Social capital does

not come from giving people information; it comes from working together with people in meaningful ways. In short, information doesn't
settle issues, people do.[8] Resolving value conflicts, which is at the heart
of the natural resource management policy process, requires trust
among people developed over time through cooperation. In greater
Yellowstone, many high-profile public conflicts (such as bison, grizzly
bear, and elk management) are "resolved" through adversarial process
rather than through cooperation and civic deliberations, often resulting in resolutions not even remotely optimal for anyone.[9] Presently,
because the GYCC's public contact is so restricted, the opportunities to
build social capital are very limited.

One typology of participation describes a wide range of possible
public involvement, from passive involvement, through consultative,
contractual, collaborative involvement, and participation "among colleagues," finally to community self-mobilization.[10] The GYCC's interactions with professionals outside the agencies and with the public are
generally of the passive or, at best, consultative types. In passive participation people are informed of projects that will be carried out or
have already been carried out. In consultative participation, people ex-
press local opinions and desires, which may or may not be taken into
consideration by officials. Public involvement in the winter visitor use
assessment and in grizzly bear management are examples.

In contractual arrangements, an agency contracts with an outside
party to do specified work with some broader impact. Some contractual work does take place through the GYCC, although very little. There
is little of the higher orders of cooperation, such as close, prolonged
collaboration, where efforts or ideas originate outside the agencies,
and people (including the agencies, NGOs, and the public) diagnose
and evaluate problems and initiatives together. Active participation
"among colleagues" focuses on empowering local people and emphasizes
activities that encourage local, informal groups to self-mobilization.
Self-mobilization is a desirable goal because it permits people in a
community to undertake their own problem definitions and solutions.
Self-mobilizing communities can, in turn, engage the agencies through
challenge projects and in other ways. There are a few associations of
this type in the Yellowstone arena, and they are growing in number
(such as citizens groups and some NGOs), but overall the GYCC under-
utilizes the various forms of cooperation available to build alliances
with diverse publics and thereby create social capital.

In the final analysis, for the GYCC to be most successful, it must get
and hold the support of the public. Not only in greater Yellowstone

and the United States, but worldwide, people increasingly demand to participate in formerly closed and exclusive government processes.[11] Everywhere people are calling for a system of government operation that is free, open, and democratic, one that encourages more deliberative civic engagement and moves beyond the adversarial relationships created by past arrangements.[12] When excluded from meaningful, genuine, effective participation, frustrated citizens resort to lawsuits, administrative appeals, and other means to force their views into the public dialogue. The true collaboration that does take place is typically both ephemeral and specific to sites and resource issues.

Public support can be built and earned through genuine cooperation and successful demonstrations of best practices. Unfortunately, the GYCC forum does not currently provide for that. There is not even a formal mechanism in place inviting the public to enter its concerns to the organization in writing. Over the two or so meetings annually, a total of perhaps an hour is provided for interacting with the public—this is not a creditable commitment for solicitation of public participation. As one researcher writes, such "tokenism" in public participation "may meet the letter of the law but is generally not helpful."[13] The GYCC is not creating and banking social capital, the ultimate policy resource available to leaders, creation of which should be a driving force behind every governing organization.[14]

In sum, the GYCC's cooperation with others is highly selective. Even on the most basic level, the GYCC's information gathering and sharing function is insufficiently thought out and is not tailored, as it should be, to support higher forms of cooperation, either site specific or regionwide. Even though the committee looks to improved intelligence as the solution to many of its problems, its meetings do not maximize the information capacity of the data and therefore cooperation falls far short of its potential. The record of the GYCC—the number, duration, and content of GYCC meetings, including patterns of cooperation with knowledgeable people outside the agencies and in the public—shows an underattention to the information content and flows needed to achieve its formal goals. The most significant goal that the GYCC has taken on for itself—to bring about ecosystem/transboundary management policy—is one of such magnitude and complexity (involving numerous interdependencies and nonroutine tasks) that it creates uncertainty for the unit managers, and in situations of great uncertainty it is even more important to encourage information input, processing, and dissemination at the highest possible levels.

Cooperation—Decision Making

Another way to examine the nature of cooperation in the GYCC is to look at its decision-making process. Choosing courses of action is one of the chief functions of leadership. It is an opportunity to achieve goals of mutual interest. Presently, the committee participates in many different, but interrelated, decision processes, both as a collective entity and as individual unit mangers. Most of the committee's decision activities, especially the estimation or intelligence function, take place on a highly selective, case-by-case basis. The committee makes decisions when it sets up agendas—cases to be attended to and people to be heard from and those to be rejected. It also makes decisions when it allocates money for projects, and it makes decisions when it terminates one project and moves on to others. After the members return from the meetings to their home units, and through communications with each other and other people, the GYCC's decision-making process continues to influence them and regional management policy. In turn, their decisions as a group are influenced by decision making within their home units, in the region, nationally, and internationally by other people and groups and in other ways. Consequently, it is inaccurate to say, as some do, that the committee does not make decisions. It is more precise to say that, although the GYCC does not make formal, binding decisions across all management units and all agencies in the greater Yellowstone region, it does make many important decisions or choices that influence management policy. How these decision-making processes are undertaken and understood is significant to the goal of transboundary management policy and transitioning toward sustainability.

Decision-Making Process

How does the committee make decisions? Does the GYCC address the full range of activities or functions of the decision process in each of the cases that it takes on? What standards are evident in its decision making? What is the pattern of the GYCC's conduct?

Collectively and individually, the GYCC does carry out most of the activities of the decision process at one time or another, but neither systematically nor explicitly. Instead, it typically focuses its decision-making activities on ordinary cases in an unsystematic, conventional way. As a result, whatever decision activities are discussed, the discussions often do not result in a comprehensive picture of the problem, its status relative to the decision process context at hand, or a program of action. This makes cooperative problem solving, communication, and joint action much more difficult. A more organized, systematic, grounded (evidence-based) approach would be to conduct a problem-oriented, contextual examination for each case, attending simultaneously not only to ordinary challenges but also to the inherent governance and constitutive challenges and implications.

Conceiving of management problems as having a "life history" is an invaluable way to understand the decision-making process. Garry Brewer described the six activities or functions of the decision process (see chapter 2) as a "life cycle."[15] Each "problem" emerges from somewhere; it is defined in size and importance; it is addressed with a strategic statement (policy); it is implemented through tactical measures (program); the corrective measures are monitored and evaluated; and, depending on results, the situation may be deemed satisfactory and the program ended, or the problem may worsen and the whole process may then be repeated to address the new larger problem. The real value of this conception of decision making is that it helps people orient to problems, their full contexts, and their solutions. It can also greatly increase the manageability of decision making. This concept, along with knowledge of what happens in each activity, awareness of the standards for performance of each function, and knowledge of common weaknesses to be avoided, all confer great advantages to users of this approach.[16]

Notes from GYCC meetings serve to illustrate this point. In the October 1996 meeting, for example, the first issue discussed was a history of the GYCC itself and its accomplishments,[17] which was followed by a wide-ranging discussion in which many people participated. Themes (problems) that emerged were that the agencies and, by implication, the GYCC needed better trust relations with communities, better organization, more coordination and more money, and crisis avoidance. The majority view taken on these matters was a top-down, bureaucratic one. The group concluded that more science and information and more money resources were key to agency leaders solving "problems." The limited discussion about the public and trust relations revealed that both governance and constitutive process were highly problematic and not fully appreciated; yet these aspects of the problem were not brought forward for explicit discussion. Like most people, the GYCC members are most comfortable discussing issues they know a lot about, can easily contain or circumscribe in concrete ways, and have the language and skills to do something about. Data in the meetings were mostly limited to presentations of historical trend information using conventional assumptions and terms. The accepted standards of adequacy employed by the group were used implicitly and remained unspoken and in the background. There was a lot of self-promotion by the committee of its present efforts and of the need to improve intelligence (that is, more biophysical, scientific management). From a promotional, self-interested point of view, the meeting was a success because the narrative or discourse remained within these restricted parameters, but the opportunity for a fuller discussion of substance and process was lost.

Another example helps illustrate concerns about the decision process at the GYCC. The October 1997 meeting examined eight management issues (sanitation at county levels relative to grizzly bear management, county planning in general, winter use in general, winter use in Grand Teton and Yellowstone National Parks, forest planning, transportation, search and rescue, and residential development appraisal).[18] Additionally, the grizzly bear program was introduced and quickly dismissed, being judged as on track and requiring no discussion (an appraisal activity that was not explicitly recognized as such). Ten people sat at the official table in the front of the room and fifteen people were in the audience. A contingent from West Yellowstone spoke up about the winter use and snowmobile issue, claiming that they felt "shut out" of the process. They wanted reliable information about winter use issues and felt they were not getting it. Representatives from the state of Wyoming also said they felt "shut out" of the winter use decision process, although this was more a demand for power from a sister agency than a demand for respect and information, which the citizens' group sought. A shared problem that became evident during the meeting was the lack of trust or social capital in the decision process. Many in the audience said they did not feel respected, nor was the meeting itself sufficiently "inclusive." The audience called for a "partnership" with the federal government to address problems of mutual concern. The GYCC responded that it had assumed that a partnership already existed (obviously an invalid assumption, given the audience's demands). It was very clear that citizens from West Yellowstone and officials from Wyoming sought to get a solid sense about what the GYCC and the federal government might decide and do. It is important for people to have stable, predictable expectations about how the decision process will unfold. With expectations stabilized about what might happen in the future from government decision making, these people could then plan their own actions accordingly. Throughout the meeting the underlying claims and counterclaims being made by people were primarily for more respect, knowledge, and power. Their value expectations and demands went unmet, as the GYCC remained unresponsive to these calls for more co-operation. In the end, as change occurs, some people will lose values and the disaffected groups will typically blame the decision-making process.

The GYCC pattern of conduct at these meetings was clear. As each management case was presented, conversations ranged across several decision process activities in a hit-or-miss fashion. One member would talk about a trend in estimation, another would speak up about an evaluative matter, and someone else would focus on goal clarification, producing a procedurally and substantively unsystematic conversation. For most cases, only a few decision functions and problem-oriented activities were attended to. Context in general received too little explicit attention. Over my twelve-year sample (1995 through 2004 and as I continued to follow events in 2005 and 2006), there was no issue taken up by the GYCC in which all decision process activities were addressed in a procedurally and substantively logical, comprehensive fashion. Instead, for all issues, the emphasis was on discussion by midlevel agency staffers who were already competent in and attuned to both technical and practical senses for the issues presented, and therefore they failed to make any real forward progress on governance or constitutive aspects of those issues.

In the second day of the October 1997 meeting, conversation among GYCC members focused on self-promotion. It was said repeatedly by the members that the committee was a good idea, that it was working, and that it addressed many important topics. The group concluded that the GYCC was a successful means to bring in experts and engage the public. The chairperson said there was not time for an adequate discussion of those points, however, and when the audience wanted to talk about improved cooperation, she repeated that there was no time and that the meeting must move on. Few handouts were available about any of the subjects addressed in the meeting. My one-on-one discussion with attendees from the public revealed that they felt that the GYCC was not a genuine problem-solving forum, that GYCC members did not really engage the issues, that the committee was just "going through the motions," and that ideas from the audience were not seriously considered, discussed, or followed up on by the GYCC. They also felt that the committee made no serious commitment to interact with the public. The GYCC was viewed as patronizing, superficial, disrespectful, and even insulting to some in the audience. Overall, the GYCC seems to function with itself and with the public as a self-promotional exchange for the agencies rather than as a genuine problem-solving entity. The GYCC chairperson consistently said that it was important to be "positive," all the while ignoring or dismissing valid audience concerns. In fact, at another meeting two years later, another GYCC chairperson said publicly that most meetings were "a waste of time and unimaginative." Despite this provocative comment, subsequent meetings up to the present have followed the same basic pattern.

Standards

What standards does the GYCC use in its decision-making deliberations, and are they of sufficient quality to ensure sound leadership and a successful transition to ecosystem/transboundary management? It is unclear whether the GYCC has explicit standards by which it assesses its own decision-making behavior. None have ever been advanced explicitly for the audience or the larger public to scrutinize, nor do they appear clearly evident in deliberations. The implicit standards the committee uses appear to be those common in conventional, everyday problem solving and those required in administration of bureaucratic agencies, including sufficing. Among these everyday standards are confusing trends with problems and rushing to the preferred management action without fully understanding first what the problem is. Such standards work in some situations but are insufficient to meet the larger needs of the committee. Transitioning to sustainability, including the basic and higher-order tasks required, requires different standards than those used day to day in routine operations.

In contrast, one of the reasons that the decision process model described earlier is so useful is that it comes with recommended standards, open to examination, and broadly applicable to all decision activities in all situations.[19] These standards are based on extensive human experience, and they support the overriding goal of finding outcomes that serve the common good. They can be used to guide the GYCC's deliberations, ensure that each decision activity is carried out as fully and effectively as possible, and avoid common pitfalls as much as possible.

Among the standards that apply to all functions of the decision process are dependability (statements of fact must be reliable, realistic, and available to everyone), comprehensiveness (all relevant data—including historical trends, conditions, and causes—and projections about the future must be thoroughly considered), and selectivity (within a comprehensive context, a specific problem and the people affected by it must be targeted). Decision making overall must be creative, open, rational, integrative, timely, nonprovocative, realistic, uniform, independent, balanced, and ameliorative. It must also create stability of expectations and show continuity. Still other standards apply to the overall decision process: Is decision making cost-effective and technically efficient? Does it have a reputation for honesty and integrity that inspires and deserves public confidence? Does it attract high-quality people and "reinvest" people so they want to participate in future decision making? Is it flexible and realistic? Is it self-aware, deliberate, and broad in scope (so that the whole decision process does not seem to be piecemeal, opportunistic, or expedient)? These are the standards that, when they are approximated in practice, allow a democracy to work.

There are also recommended standards for making sure that decision making effectively serves common interests.[20] All these standards apply to ordinary, governance, and constitutive challenges although in different ways. How effectively do decision makers arrange for common interests to prevail over special interests? Do they constantly give precedence to high-priority over low-priority common interests? Do they protect both the interest of the community and also the interests of individuals at the same time? When there are conflicting individual or exclusive common interests (such as property), do they give preference to participants whose value position is most substantially involved? Do they allocate authority and other values (for example, power, money, knowledge, skill, and so on) to ensure that the most important priorities are achieved? These are all important considerations for leaders.

It is clear that the GYCC does not explicitly, consciously, or systematically use these standards, although, like any group of intelligent leaders, it does gravitate to many of them. There is no discussion at committee meetings about what standards to use or the adequacy of those currently in use. As a result, many low-order ordinary cases are on the agenda along with genuinely higher-order issues, and all are treated more or less equally, with no effort to sort out these issues of greatly differing import in terms of the common interest. For example, the likelihood that lynx might be listed under the Endangered Species Act at the time it was discussed has implications for agency management, to be sure, but it is not as important as the creation of a system of data acquisition, processing, and management (that is, organizing an overall intelligence activity) for the entire greater ecosystem. At present, the GYCC's deliberations involve cases that are a mix of low-, mid-, and high-order common interests. Additionally, some cases involve inclusive (important to everyone) common interests (such as fire policy), whereas others involve exclusive (important to a few people) common interests (such as weed surveys or a specific sign for public education). No distinction is made among these, and all are given more or less equal time and attention. Consequently, the committee's decision-making activities overall favor the status quo and the present style of decision making and tend to advance the committee's unofficial goals.

Pitfalls

Certain pitfalls or weaknesses plague each of the activities in the typical decision process.[21] During initiation, a common pitfall is that the initial problem definition fails to capture the full and true nature of the emerging problem. In estimation, analysis of the problem and gathering of intelligence may be inadequate. During selection, poor coordination in government decision making may lead to more bureaucracy and gridlock. Pitfalls in implementation include poor coordination, bureaucratic overcontrol, exclusion of key parties, or delays. Among the pitfalls in evaluation are insensitivity to criticism, especially from outside the agencies, or failure to learn from experience. And for termination, pitfalls include pressure to continue unsuccessful policies and programs or failure to prepare for termination early enough in the policy process.

Do the GYCC's decision-making practices permit it to recognize and avoid these and other common pitfalls? Three examples give us insight. First, by its own admission, the "Vision exercise" did not produce the intended consequences for the committee. It was, among other things, an intelligence or estimation failure.[22] Moreover, the GYCC learned the wrong thing from this experience. Instead of learning and developing its capacity to implement a fully problem-oriented, contextual approach to strategy development, the GYCC learned only to keep its head down and avoid similar initiatives. It never did go back and examine its original operating assumptions as a basis for learning and improvement. Second, the congressional hearings of the mid-1980s on government coordination in greater Yellowstone were the product of delays and other common pitfalls. The GYCC had existed for more than twenty years at that time with a primary goal being coordination among participating units, yet the hearings revealed that little coordination had been done despite the availability of resources to do so.[23] A timely, comprehensive, and reliable coordination effort early on might well have avoided the negative feedback and bad public relations the agencies received for the committee's well-intentioned but less-than-effective efforts during its first twenty years of existence. Since that time, coordination has been increased. Third, my observations and discussions with GYCC members, as well as newspaper reports, show that the committee and the agencies in general are relatively insensitive to criticism, even constructive, valid criticisms.

This seems to show an inability or unwillingness to learn and change. It also speaks to the power of leaders and the agencies in that they can, with impunity, ignore legitimate concerns from the public and from their own staffs. Taken together, these pitfalls cause the GYCC to fail to recognize and capitalize on opportunities for progress. Institutionalizing an open, inclusive, and comprehensive approach, that is, an active learning system, could have avoided many such pitfalls evident throughout the GYCC's efforts. Taken together, the committee's inability to avoid these pitfalls diminishes its leadership potential and its ability to move toward sustainability.

Implications

What can be said of the committee's approach to decision making? The pattern of conduct over the years, independent of the committee members who come and go, is clear. The committee's approach appears to be deeply rooted in a commitment to technocratic problem solving, or scientific management, buried in agency bureaucratic structures and cultures and in a defensive policy stance taken by leaders. Although coming together to address problems is beneficial for technical, political, and other reasons, the conventional approach taken by the GYCC overemphasizes ordinary problems and deals with them incompletely. A technical (and biophysical) focus is taken on the ordinary cases, presumably in an attempt to gain substantive rationality (that is, to get and master the ecological "facts"). The decision-making process that is being used and its adequacy are never formally or openly appraised. Thus, the quality of procedural rationality in deliberations is little considered and is never systematically appraised or sought after. Recommended standards of problem orientation and decision making that would help encourage quality team performance and high-quality leadership are never used explicitly. As a result, conventional decision-making pitfalls abound. The GYCC typically avoids or neglects analysis of governance and constitutive challenges, probably because they are viewed as "politics" to be avoided. This pattern of conduct has several implications.

When the GYCC operates in this mode, it is too selective; its focus of attention is too restrictive when it comes to the decision process. For example, estimation (or intelligence) activities dominate the focus of attention, while the most neglected of the decision process activities is termination. It has been suggested by Garry Brewer and others that the best way forward to achieve ecosystem management is to treat it as a termination problem.[24] In other words, current nonsustainable practices must be terminated. Identifying these harmful practices is the first job, and the second is to find effective yet equitable ways to end them. In one sense, this is the job of the GYCC.

Ideally, the decision process balances selective problem solving with comprehensive problem solving. Careful analysis and discussion must range from one to the other. In this way, problem solvers make sure they simultaneously gain a specific targeted picture of the problem before them as well as a thorough appreciation of the larger contextual setting. The GYCC's tendency toward too narrow a focus leads to an incomplete picture of the problems and their interconnectedness, in turn encouraging a "disjunct incrementalistic" approach to problem solving.[25] Members fail to recognize fully that the ongoing decision process is really about humans making value choices, regulating their own practices through institutions that use and affect natural resources.

There is no evidence that the GYCC actively, consciously strives to meet the standards for serving common interests or for ensuring that the decision process for each management issue is carried out well. The standards presented above, applied consistently, are the only way to create a process that serves the common interest. Common interests are served when the decision-making process is inclusive and open to broad participation, when it meets the valid expectations of participants, and when it adjusts actions as decisions are implemented so as to be responsive and adaptive in achieving goals as the context changes. Attending to these standards could significantly aid the committee's challenging leadership job.

The GYCC does not distinguish between different kinds or levels of problems—ordinary, governance, and constitutive challenges, their associated decision processes, and their interconnected consequences. Because of its position as a cooperative federal body and chief custodian of the public trust, the GYCC should be important in the region's constitutive policy process. Helping society adjust the constitutive process to meet changing realities of our time should be a central task for leaders. Even though GYCC is not the only level that can help lead adjustments, it is nevertheless very well situated to do so. Ideally, higher-level officials in the agencies and appointed officials could also be very helpful to the public and GYCC, if they chose to help. If the committee fails to appreciate this fact and act accordingly, it then serves in fact to maintain a status quo policy. It is not possible to transition to sustainability by supporting the status quo. Even though the GYCC may not be the optimal level of government for making governance and constitutive adjustments, it is ideally situated to understand the issues and their contexts and to serve as a catalyst for its agencies,

departments, and Congress to become active in promoting change toward sustainability. Decision processes, both governance and constitutive, address the value demands of the community on important
issues, on matters of process as well as substance. These decisions determine how natural and cultural resources are allocated and developed, how wealth is produced and distributed, how human rights
are promoted or diminished, how enlightenment is encouraged or
retarded, how health is fostered or neglected, how rectitude and civic
responsibility are nurtured or blighted, and so on through the whole
gamut of demanded values. Making these distinctions in actual cases
is an important part of successful leadership.

In sum, the current level of the GYCC's cooperation in meetings and
with the public through decision-making processes is disappointing,
given its tremendous potential. Cooperation can be measured in terms
of its scope (how widely shared goals, values, and beliefs are), strength
(how authority, knowledge, and position are distributed), and duration
(how much staying power it has).[26] Overall, the GYCC's cooperation has
taken a weak form given its formal goals, its aspirations for inclusivity,
and the diversity of people and publics it could affect. The committee's
pattern of selective decision making, focusing on ordinary problems in a
case-by-case way and emphasizing the estimation activity to the neglect
of other decision functions, makes the GYCC highly susceptible to decision process pitfalls and failures. It also traps the committee in conventional, scientific management and in bureaucracy. Widely recognized
standards to promote common interest outcomes are either not known
or not met explicitly and actively. This, in turn, limits problem-solving
capacity and as well as cooperative potential. This pattern of conduct
blocks active learning, maintains and reinforces current outlooks, and
promotes status quo ways of cooperating and decision making.

Demonstrations—Activities

For true ecosystem/transboundary management policy in the greater
Yellowstone region to become reality, the region's leaders must not
only be successful in problem solving and cooperating, but they must
also demonstrate to others what they want to achieve across the
region and what ecosystem/transboundary management means in
practice.[27] They must show the value-added benefits of any new policy
for managing natural resources. One widely recognized way to bring
about constructive change is to create demonstration projects or programs that will illustrate or explain how the new system will operate.
These exemplars or exhibits, which can be at small or large scales, can

serve as tools for active learning by those who create them. In turn, they can be observed as best practices, studied by everyone, evaluated, and adapted. Progress toward ecosystem/transboundary management policy will most likely be built on a track record of successful demonstrations, which, in fact, is how most people learn new practices.

The Committee

We can look at the GYCC's track record of practical demonstrations as including not only experiments in the field but also innovative changes in the functioning of the committee itself. Is the committee a self-created best practice? Has the committee set itself up and managed its own operations to be a demonstration or a new model of world-class problem solving, cooperation and coordination, and integrated leadership to foster sustainability? Can the GYCC's meetings be taken as a demonstration of what the committee is trying to achieve? Does it represent the effective problem solving, cooperation, and integrated leadership that are needed to transition the people, organizations, and institutions in the region to ecosystem/transboundary management policy?

Milestones

The GYCC takes pride in its accomplishments. The committee lists "milestones," but it is unclear if these are being promoted as accomplishments or whether they are just efforts in progress with future expected payoffs. Either way, a number of these constitute demonstrations of one kind or another, including its outfitter policy, the "Aggregation" exercise, its now inactive Science Committee, and its GIS/data work. These and other works between 1964 and 2000 were highlighted in the committee's only public brochure about its work (table 4.1, see Web site http://www.gycc.org).[28] We can examine these to see whether they are accomplishments of real significance. What might the GYCC reasonably be expected to have accomplished over the last four decades or even in the last decade, given the authority, goals, and resources it has had at its disposal? Is the GYCC's list of milestones a creditable track record? Has it moved us toward sustainability through ecosystem/transboundary management policy in significant ways?

The GYCC's mere existence is a valuable asset to the region, but its track record of accomplishment is not up to the standards it has set for itself. Some milestones are clearly a product of the GYCC's efforts, namely, the "Aggregation" and the winter use assessment. But in other

Table 4.1. "Milestones" or Accomplishments Published by the Greater Yellowstone Coordinating Committee

1964	Greater Yellowstone Coordinating Committee formed with signing of a memorandum of understanding between the National Park Service and Forest Service.
1979	GYCC issues "Guidelines for Management Involving Grizzly Bears in the Greater Yellowstone Area."
1983	The "Bald Eagle Management Plan" released, covering five GYCC units.
1985	Joint hearings held by House Subcommittee on Public Lands and National Parks and Recreation about coordinated management in the GYA.
	A joint Forest Service/Park Service planning team established in Billings to create an information base from existing planning documents.
1986	Memorandum of understanding between Park Service and Forest Service revised to reinforce existing mutual cooperation and coordination in response to congressional hearings.
1987	*Greater Yellowstone Area Aggregation of National Park and National Forest Management Plans* released. The report compiles and summarizes existing management plans for the national parks and forests within the GYA.
1988	"Greater Yellowstone Area Interagency Fire Planning and Coordination Guide" completed.
1989	"The Greater Yellowstone Postfire Assessment," a collection and evaluation of postfire data compiled by fifteen interagency teams, is published.
1990	Draft "Vision for the Future" document released for public comment. The "Vision" document describes a desired future condition for the GYA.
1991	GYCC issues a "Framework for Coordination of National Parks and National Forests in the Greater Yellowstone Area." The "Framework," a final version of the 1990 "Vision" document, includes guidelines and principles for coordinated management of the GYA.
1992	"Guidelines for Coordinated Management of Noxious Weeds" released. The document served as a model for coordinated and integrated noxious weed management.
	National Forest Service issues GYA outfitter policy to provide consistent direction for the administration of outfitter guides.
1993	Units of the coordinating committee issue wilderness fire management plans for wilderness and backcountry areas.
	National Forest Service issues special orders on the use of weed-free feed to reduce the spread of noxious weeds.
1994	National Forest Service issues uniform regulations on wilderness and nonwilderness recreation use.
	GYCC forms Winter Use Management Work Group to analyze current winter use patterns and areas of conflict.

(continued)

Table 4.1. (*continued*)

1996 GYCC funds provided for completion of the grizzly bear cumulative effects model.

1999 "Winter Visitor Use Management: A Multi-Agency Assessment" released.

"Greater Yellowstone Area Air Quality Assessment" released.

"Effects of Winter Recreation on Wildlife of the Greater Yellowstone Area: A Literature Review and Assessment" published.

2000 Initiated national forest winter use monitoring.

Development of GYCC Project Program—thirty projects receive GYCC grants as they advance GYCC conservation priorities and represent new partnerships, methods, or information.

2001 "Watershed Management Strategy for the GYA" released.

2002 "Weed Pocket Guide" released.

2003 Best management practices clarified.

2004 Stewardship projects report.

Coordination of four national forest planning efforts.

2005 Initiation of Biofuels, Recycling, and Bulk Purchase Subcommittee.

Source: Greater Yellowstone Coordinating Committee, "GYCC at Work," in *Greater Yellowstone Coordinating Committee Briefing Guide* (Billings, Mont.: Greater Yellowstone Coordinating Committee, 2001), 9, and more recent additions.

cases, the somewhat disjointed record suggests that the GYCC cast its net widely and claimed whatever it caught as evidence of its own accomplishments. (Many groups have a tendency to claim credit for the accomplishments of others who may be peripheral or tangential to the organization; this speaks to the importance of legitimacy and the struggle to achieve it.) In some instances, the accomplishments cited are actually those of independent committees (for example, Tri-State Trumpeter Swan Group and Jackson Hole Elk Working Group). The committee even lists a 1979 grizzly bear publication that helped standardize bear management, which was a product of a graduate thesis by Steve Mealey, who largely did the work and publication on his own and later went on to become forest supervisor of Shoshone National Forest.[29] This publication clearly helps unit managers to manage their respective agency activities relative to grizzly bear habitat, but it is a stretch to say that the GYCC initiated or oversaw the research and publication. Other milestones that the GYCC lists and takes credit for (even if implicitly) were started, and operate today, independently of the GYCC. The Bald Eagle Working Group is one example. This group, which issued the

1983 "Bald Eagle Management Plan" and continues its work today, was set up by professionals in and out of the agencies in the ecosystem. To be sure, these committees represent the growing organization of the arena and the growing focus and coordination throughout greater Yellowstone, a phenomenon that has multiple origins, most of which are independent of the GYCC. In contrast, the GYCC deserves credit for attempting to bring these disparate activities under its tent. Today, the GYCC's executive secretary stays in touch with regional committees and subcommittees and their work, informally at least.

It is debatable whether this record of achievements demonstrates a genuine and substantive commitment to transboundary management policy, enhanced coordination, and improved integration and whether it demonstrates the very best practices that GYCC is capable of producing. It is not always easy to distinguish between real accomplishment and self-promotion in an organization's materials and meetings. Nevertheless, it is reasonable to expect that, given nearly forty years of federal effort, the GYCC's record would be much stronger. Nevertheless, it still has potential to achieve significant successes. The mere existence of the committee demonstrates that the agencies recognize a need for improved government behavior. The GYCC carries a high degree of authority signature (official high-level status) and control intent (leaders make clear that they intend to make changes) behind its management policies, both of which are very important to future change and adaptation.

The "Aggregation" and "Vision" Exercises

Other measures of the GYCC's demonstrations exist, including several high-profile efforts actually initiated and implemented under the committee's auspices, specifically the "Aggregation" and "Vision" exercises. Both of these followed congressional hearings in the 1980s, which showed that the Park Service, Forest Service, and Fish and Wildlife Service were not as effective as needed in their efforts to coordinate and integrate data and management across boundaries.

In 1985 and 1986, the U.S. Congress, with the help of the Congressional Research Service (CRS), launched an in-depth study of federal land management. The study produced the 1986 publication *Greater Yellowstone Ecosystem: An Analysis of Data Submitted by Federal and State Agencies*. The major finding of this study was that existing agency information was inadequate to analyze the site-specific impacts of proposed actions or to resolve management conflicts in greater Yellowstone. Where data were available, measurement standards used by different agencies were often incompatible. For example, the CRS said that both the Forest Service and Bureau of Land Management generally used "animal unit months" for grazing leases and permits, but that Beaverhead, Caribou, and Targhee National Forests reported sheep grazing in "numbers of animals." In addition, federal agency data on noncommodity resources, such as recreation, wildlife, and cultural resources, were incomplete, not maintained over time, and not very site specific. This CRS finding spotlighted the inadequacy of coordinated management of federal lands in the GYE. Combined with increased public interest, the CRS report spurred the Forest Service and Park Service to reexamine their data management. At the time, the agencies were quite concerned that Congress might dictate policy and force the agencies to change to meet new goals, which would limit agency discretion.

In response, the two agencies undertook a new project under the auspices of the GYCC.[30] In 1987 the GYCC published its report, *The Greater Yellowstone Area: An Aggregation of National Park and National Forest Management Plans.*[31] This report described, for the first time, existing natural resources and human use patterns in the Greater Yellowstone Area to the extent that these were known. The principal objective was to compile information on the relation between the parks and forests and provide an overview of their management. The "Aggregation," as this document is generally called, underscored the disharmonious nature of management and mapping conventions among the agencies and clearly illustrated the need to find a common focus for the management of regionwide resources and to establish an administrative structure and process to resolve conflicts and set positive direction. This document is, as its name suggests, no more than a compilation or collection of plans and data without any significant or detailed integration, evaluation, or conclusions. Although a helpful demonstration, it cannot be taken as a serious effort at integrating or evaluating data gathering, processing, analysis, or dissemination and thus is not a useful learning exercise.[32] It does function, however, to show managers on individual units how their areas fit into a larger geographic context and has value for that reason.

The "Vision" exercise turned out to be much more important than the "Aggregation." Again in response to the CRS report, the GYCC in 1989 appointed four staff members from the Forest Service and four from the Park Service to begin work on an overarching mission statement—a "vision" to guide coordinated interagency management. The agencies and the GYCC expected that this effort would provide a common focus for setting goals and implementing them. Both forest and park plans could be amended to meet the new vision. Loraine Mintzmyer, then regional director of the National Park Service in Denver, said, "To meet the congressional expectation of prospective review and analysis, an interagency document was anticipated—one which would describe the future condition of the greater Yellowstone area through coordinated management goals and how they could be achieved. . . . This was not simply to be a regional plan or decision document—it was to be a study of the conditions for the areas involved, a recognition of goals, and a formalization of coordinated, guiding principles. This document was to be a model for interagency cooperation in this area and a model for other areas, well into the next century."[33]

The GYCC released the seventy-four-page draft "Vision" document in 1990.[34] The goal of the exercise was stated clearly in the draft document: "The first step to accomplish these objectives is the creation of an interagency document to describe the desired future condition of the GYA through coordinated management goals and how they can be achieved. The Vision provides this description and sets the stage to complete applicable plan amendments, if needed."[35] Three overriding "principles" on which to ground management were to conserve the sense of naturalness and maintain ecosystem integrity, to encourage opportunities that were biologically and economically sustainable, and to improve coordination.

The "Vision" was an expensive three-year effort to address problems outlined by the CRS. It was the GYCC's response to the CRS evaluation. The "Vision" document was circulated widely in greater Yellowstone for comment, and the comments that came back were numerous and fierce.[36] Many interest groups and individuals responded to the draft, nearly all of them negative. Both the environmental community and the commodity and recreation communities opposed it because, presumably, it did not meet their expectations. The governors of Wyoming, Montana, and Idaho asked that it be withdrawn. Newspapers were full of articles in strident opposition.

The volume and vehemence of negative responses shocked the GYCC and others. Redrafting of the document was taken away from the original authors and assumed by the GYCC itself. The final version was reduced to an eleven-page brochure, renamed the "Framework," and most of the major points in the "Vision" were abandoned. Ed Lewis, executive director of the Greater Yellowstone Coalition at the time, subsequently said, "The process to create a visionary management strategy has been gutted. These agencies spent three years developing a plan that calls for business as usual."[37] He also said that the "Framework" "is a prescription for continuation of business as usual. It allows development activities such as logging, oil and gas and mining to continue essentially unabated on Greater Yellowstone's public lands. Most disturbing, the final document, now called 'A Framework for Coordination of National Parks and

National Forests in the Greater Yellowstone Area,' completely abandons the concept of ecosystem management which was at the heart of the draft Vision released a year earlier."[38]

Colleague Pam Lichtman and I examined four widely circulated explanations for the failure of the "Vision" exercise—"unclear objectives," "a politicized environment," "miscalculation of public response," and "sinister agency conspiracy"—each espoused by one or more interest groups.[39] We concluded, using the problem-oriented, contextual, and decision-making standards discussed previously, that the GYCC and the agencies had failed for several reasons. First, the largely midlevel forest and park agency staff members who had drafted the document were perhaps not the best choice to undertake a task with such serious political ramifications. Nor were they given enough time and resources to orient to the problem adequately, assuming they knew how to do this. Second, they did not carry out a thorough contextual analysis, that is, they did not assess the constituencies, external political environment, regional economic forces, and their agencies' own policy preferences and biases. Third, the analytic method used was limited and conventional. It was not fully problem oriented, contextual, and multimethod. For example, the basic problem was not identified as a constitutive or even governance problem. If it had been, then the solution/document would have contained different material, and it would have required a different methodology than what was used. Although these are common pitfalls known to plague bureaucracies, it seems that no special effort was made to avoid repeating them. There are lessons from both the "Vision" and "Aggregation" exercises that went unlearned. Both incidents provide invaluable insights into the workings of the GYCC, especially about how it blocks its own capacity to learn.[40]

Implications

Appraisals can be done by comparing the committee's demonstrations with what people outside the committee say, with recognized standards, and with what might reasonably be expected of the committee, given its goals, authority, money, staff, and time available.

Although it is essential to have an interagency and interunit committee as a demonstration that meets at least semiannually, the GYCC's actual deliberations function largely as information reporting or information exchange about management matters of shared interest. Committee members also learn about issues of importance in the region and what other unit managers are doing. There is, however, minimal analysis, discussion, and solution of overall problems. These meetings largely add to the estimation or intelligence activity of the decision process and the trend-mapping task of problem orientation. But despite the opportunities they offer, they fall short of being fully problem-oriented, contextual, multimethod, problem-solving demonstrations. They also serve social functions as value-shaping and -sharing exercises: they permit managers to check and align their expectations with one another, and they provide opportunities for rituals of mutual respect and affection (the committee is much more a support group than it is a problem-solving group). Overall, the GYCC's meetings have not been a good demonstration of best practices of problem solving for sustainability. In early decades of the GYCC's work, the record of demonstrations was thin, but in recent years it has grown as issues have proliferated, forced their way into the focus of attention, and found a place on the agenda. Although the committee's reports and publications have been of real value, they have not been produced through any overall systematic or strategic approach. In some instances, what is represented as progress has been responses to "crises," such as the congressional hearings, fires, or legal challenges. Thus, the GYCC is being pushed by circumstances rather than leading in the vanguard toward improved management policy for sustainability. The committee functions in part as a defensive coalition that supports its members and their collective operations against challengers and critics, rather than leading transboundary management policy into existence. As such it is not a good demonstration of what is needed. Although the GYCC itself has not been a strong demonstration of a best practice, the staffs of several members' units are in fact doing some excellent demonstrations, most of which operate under the radar scope of these high-level officials.

The two high-profile demonstrations that we have discussed demonstrated that, indeed, the region was highly fragmented in terms of authority and control and that the GYCC itself was out of touch with its context. But in the end, the committee learned the wrong thing from both exercises. It learned not to stick its neck out rather than learning to further its aims in other ways using smarter, more contextual approaches or perhaps smaller initiatives that directly supported improved coordination and management policy. More recently, the GYCC has shifted much of its time and money to "safe" projects, such as soil and weed mapping. Thus, there are few solid demonstrations of best practices—what better coordination or ecosystem/transboundary management policy might look like in practice.

In sum, the committee's meetings, track record, and other demonstrations to date are helpful and laudable, but they fall far short of what is possible and what is needed—they are necessary steps but far from sufficient. The overall record of accomplishment, including those discussed below, seems weak given the human, organizational, and institutional resources at hand. In light of the GYCC's goals and the decades it has had to improve coordination and integration and bring about new management policy, it is reasonable to expect that that the GYCC could have achieved far more than the record shows.

Demonstrations—Projects, Staff, and Subcommittees

The work of GYCC is evident not only in its meetings but also in how it sets up and supports projects. These, the working part of the decision process, or implementation, require organization, committees, workers, and, of course, money and are also demonstrations. The Congressional Research Service said in 1986 that the existing interagency coordinating committees in greater Yellowstone were not comprehensive in either membership or approach, specifically noting that the GYCC, which had the broadest federal participation in the region, excluded Caribou National Forest, the Bureau of Land Management, and unspecified U.S. Fish and Wildlife Service units. Today, the Caribou Forest and the U.S. Fish and Wildlife Service, represented by the National Wildlife Refuge system, are on the committee, although the Bureau of Land Management and others are still not participants. Also, the GYCC now has special monies within its budget to use each year to advance its goals. Since 2000 the GYCC has supported projects that it feels advance its goals, each year allocating about $300,000 to fund a variety of demonstration projects. Looking at how the money is allocated and the projects that it supports gives us insight into what work the GYCC thinks is important for demonstration purposes. Also, looking at the committees, working groups, and staff that the GYCC works through or is related to gives us further insight into how it views the demonstration work that needs to be done. Again, some of these committees and working groups are GYCC initiated whereas others are fully independent. Data for this analysis are from GYCC reports and meeting minutes.[41]

Money for Projects

The GYCC uses three criteria to evaluate proposals it considers for funding. *Strategic fit* measures how well the project addresses GYCC priorities (ranked goals). Proposals are given scores of 3 (strong fit with GYCC priorities—lands, water, fish, weeds), 2 (meets second tier priorities—data management, whitebark pine, threatened and endangered species, public information/recreation), or 1 (important but not on the priority list). *Application* evaluates how well the project advances goals for the entire greater Yellowstone area versus a single unit. Again, proposals are scored as 3 (benefits that span the entire ecosystem), 2 (benefits more than one unit), or 1 (primarily benefits only one unit). Finally, *resource leveraging* evaluates how well the project incorporates partnerships and leverages resources. Proposals are scored as 3 (project funds are leveraged greater than 1:1 match with partnerships), 2 (project funds are leveraged but less than 1:1 match with internal partnerships), or 1 (project funds are not leveraged and have no internal partnerships).

Table 4.2 and appendix 3 list how the money has been spent. The 2001 GYCC Briefing Guide mentions seven project or interest areas: noxious weed management, wildlife studies and projects, soil and watershed management, Yellowstone cutthroat trout conservation efforts, whitebark pine restoration and management, land pattern studies, and recreation and visitor services.[42] From 2000 through 2004 a wide variety of projects was funded, some repeatedly. In those five years, the GYCC received more than three hundred proposals and funded nearly two hundred, totaling about $1.5 million. On balance, it appears that many of the funded projects were relatively localized, low-order projects with few regionwide, high-order impacts. In 2004, for instance, the committee received sixty-nine proposals requesting $900,000 and funded twenty-five totaling $284,000. The first-tier priorities that year were invasive species management, watershed management, land patterns, and native cutthroat trout; and the second-tier priorities were whitebark pine, data management, recreation management, and threatened, endangered, and rare species. There is clearly a large latent demand for project work that far exceeds the GYCC's dedicated resources. The GYCC Web site summarizes projects funded in priority areas.

The GYCC Web site also lists nine priority areas for 2005–6: land patterns and land uses; native cutthroat trout conservation; watershed management; invasive species prevention and management; recreation management; whitebark pine conservation; grizzly bear recovery; wolverine and lynx conservation; and data acquisition, management, and sharing. Funding patterns in 2005 and 2006 were similar. In 2005 the GYCC received sixty-one proposals of which thirty-nine were funded for a total of $296,000. Many of these reflected the nine priority areas, but other lower-priority projects were also funded, for example, a portable wash station, ATV trail use monitoring, and sand and gravel certification. In 2006 ninety proposals were received, of which thirty-eight were selected to receive a total of $270,000. The selected projects brought with them $1,231,200 in partnership funds and in-kind contributions to the region for the advancement of the eight GYCC priorities.

Table 4.2. Funds Allocated by the Greater Yellowstone Coordinating Committee in support of projects in the Greater Yellowstone Area, 2000–2004.

Federal Management Units	2000	2001	2002	2003	2004
Beaverhead-Deer Lodge National Forest (BDLNF), Montana	2 projects, $15,000	3 projects, $17,000	2 projects, $17,000	3 projects, $21,000	3 projects, $15,000
Bridger-Teton National Forest (BTNF), Wyoming	3 projects, $28,000	5 projects, $29,000	5 projects (1 with GTNP—$2,500), $41,500	8 projects (1 with SNF—$12,500), $51,500	5 projects, $30,000
Custer National Forest (CNF), Montana	2 projects, $15,000	3 projects $22,000	3 projects (1 with GNF—$3,000), $16,000	5 projects (1 with GNF—$4,000), $25,500	2 projects, $6,000
Caribou-Targhee National Forest (CTNF), Idaho	5 projects, $33,500	6 projects, $49,000	6 projects, $27,000	8 projects, $42,000	5 projects, $30,000
Gallatin National Forest (GNF), Montana	6 projects, $38,500	9 projects, $53,000	5 projects (1 with CNF—$3,000), $34,000	8 projects (1 with CNF—$4,000), $39,000	5 projects, $30,000
Shoshone National Forest (SNF), Wyoming	2 projects, $30,000	4 projects, $27,000	3 projects, $22,500	7 projects (1 with BTNF—$12,500), $49,000	4 projects, $30,000
Grand Teton National Park (GTNP), Wyoming	2 projects, $30,000	9 projects, $22,500	3 projects (1 with BTNF—$2500, 1 with NER—$5,000), $12,500	4 projects, $21,000	3 projects, $15,000
Yellowstone National Park (YNP), Wyoming	3 projects, $15,000	4 projects, $27,000	3 projects, $25,000	4 projects, $23,500	3 projects, $22,500
Greater Yellowstone Coordinating Committee, Montana	6 projects, $105,000	6 projects, $69,000	9 projects, $110,000	2 projects, $28,000	5 projects, $105,000

Table 4.2. (*continued*)

Federal Management Units	2000	2001	2002	2003	2004
National Elk Refuge (NER), Wyoming			1 project with GTNP, $5,000		1 project, $5,000
Totals	31 projects, $310,000	44 projects, $300,000	36 projects, $300,000	47 projects, $284,000	36 projects, $288,500

Most reflected GYCC priorities. There are, however, unspoken or informal rules for distributing the money. Each unit is to get some money, although those projects may not be high-order ones. Money is distributed in more or less equal sums across all units. The informal goals override formal goals, thus another example of goal displacement or inversion.

Staff and Subcommittees

The GYCC has a single dedicated staff member but draws on agency employees of the management units on a case-by-case basis for specific tasks. It also sets up its own longer-lasting subcommittees, typically staffed by agency people as well. These can be examined as part of the GYCC's record on cooperation and demonstrations or best practices.

The one full-time dedicated staff member is the GYCC's executive coordinator, who manages its activities, including analyses performed by the GYCC and development of support needed to accomplish its objectives. The two individuals, Larry Timchak and Mary Maj, who have held the position since its creation in 2000, have performed very well. The coordinator's work, as described in chapter 3, includes:

finding additional funding from donors, grants, and alternative funding;
administering the funding program, coordinating and tracking the projects of work groups, task forces, and other interagency groups, and monitoring accounts;
representing the committee at the field level, including giving presentations, attending meetings, and interacting with the media, Congress, officials, conservation groups, and others;
coordination, contact, and liaison among the federal agencies, communities, and state and local governments;
helping the committee to develop goals, strategic planning, standards, and other management mechanisms;
and monitoring and reporting.

This is an enormous amount of work for a single individual. The GYCC also draws on its unit staffs for various specific projects and for the subcommittees it sets up or authorizes, some of which have existed for years.

The 2001 GYCC Briefing Guide states that "various [sub]committees are responsible for the on-going coordination of management activities in the Greater Yellowstone Area (GYA)."[43] Figure 4.1 shows the GYCC in the center of a web of fifteen subcommittees, and table 4.3 describes the activities of each. Nine subcommittees are listed as "linked" and report directly to the GYCC, and six are "related" but do not report directly to the GYCC. Most subcommittees that report directly were started independently of the GYCC (for example, Tri-State Trumpeter Swan Group and GYA Bald Eagle Working Group). The six committees that do not report to the GYCC function independently (including the Interagency Grizzly Bear Committee, Yellowstone Ecosystem Subcommittee, and Greater Yellowstone Brucellosis Committee). Despite the figure listing fifteen subcommittees, there are few actual committees that came into existence by direct mandate of the GYCC or that function under the primary guidance of the GYCC to meet GYCC goals and priorities.

These fifteen subcommittees, representing some of the most high-profile working groups in the region, address diverse management challenges. Figure 4.1 gives the impression that centralized cooperation is more extensive than it in fact is. In addition to these groups recognized by the GYCC, there are many informal and temporary working relationships throughout greater Yellowstone that do not report to the committee and are not related to it. These are formed not only among federal staff but also include a few university and nongovernmental people.

Two recent promising groups are the agency staff members who worked on assessment of recreation in GYA, ably assisted by Susan Marsh of Bridger-Teton National Forest, and a "sustainability" group led by Anna Jones-Crabtree of Bighorn National Forest and John Allan of Yellowstone National Park. The work of these groups demonstrates the advantages of using teams well and freeing up midlevel professionals to do the regionwide assessments needed. More of this kind of monitoring and evaluation is sorely needed. Jones-Crabtree's and Allan's subcommittee offers ecosystemwide benefits by looking at biofuels, recycling, and related activities across all federal agencies.

Implications

The GYCC has an opportunity to demonstrate what it stands for and what it is seeking through the projects it pursues. It can highlight what

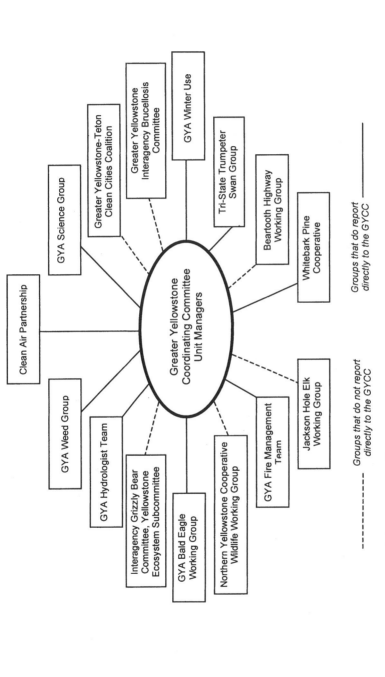

Figure 4.1. Committees of the Greater Yellowstone Coordinating Committee
Source: "Status of GYCC Committees," *Greater Yellowstone Coordinating Committee Briefing Guide* (Billings, Mont.: Greater Yellowstone Coordinating Committee, 2001), 7.

Table 4.3.   Working Groups and Other Subcommittees, Greater Yellowstone Coordinating Committee

GYA Clean Air Partnership. The committee consists of unit air resource program managers as well as the Departments of Environmental Quality in Idaho, Montana, and Wyoming, and the Idaho National Engineering and Environmental Laboratory. The committee serves as a technical advisory group on air quality issues to the GYCC and as a forum for communicating air quality information and regulatory issues, and it coordinates monitoring between state and federal agencies.

GYA Fire Management Team. Fire management officers from each GYCC unit meet each spring and fall to review fire management planning status and operational procedures. Fire managers provide peer review of individual unit fire management plans and develop procedures for coordinated management of large and or complex fire incidents within the GYA.

GYA Science Group. The group helps identify priority research needs and coordinates research projects across the GYA. Develops and analyzes scientific information to provide a scientific basis for the management of natural and cultural resources in the GYA. Currently inactive.

GYA Weed Group. Invasive species coordinators from each unit work together on common inventories, establishment of cooperative weed management areas, and integrated management to prevent the spread of noxious weeds.

GYA Hydrologist Team. The team works on a GYA-wide assessment of watershed conditions, restoration priorities, monitoring, and cooperative management opportunities.

Whitebark Pine Cooperative. Partners include forests and parks in the GYA, Wyoming Game and Fish, Forest Service Research, USGS Interagency Grizzly Bear Study Team, Forest Service tree nurseries, and the Wyoming State Forestry Division. Partners are working to maintain and restore whitebark pine stands threatened by white pine blister rust.

Tri-State Trumpeter Swan Group. State fish and wildlife departments for Idaho, Montana, and Wyoming; U.S. Fish and Wildlife Service; and other federal land managers in the GYA are working to maintain and restore trumpeter swan populations and habitat.

Northern Yellowstone Cooperative Wildlife Working Group. Biologists from Yellowstone National Park, Gallatin National Forest; USGS Biological Resource Division; and Montana Fish, Wildlife, and Parks are working together on issues concerning management of the northern elk herd and other ungulates.

Jackson Hole Elk Working Group. The National Elk Refuge, Grand Teton National Park, Bridger-Teton National Forest, and Wyoming Game and Fish Department deal with issues primarily related to elk and bison winter range management.

Greater Yellowstone Bald Eagle Working Group. Biologists from Yellowstone National Park; U.S. Fish and Wildlife Service; and the states of Wyoming, Montana, and Idaho coordinate the recovery of the bald eagle.

Yellowstone Ecosystem Subcommittee (YES). A subcommittee of the Interagency Grizzly Bear Committee, consisting of representatives from the U.S. Fish and Wildlife Service; Idaho, Montana, and Wyoming wildlife departments; the Forest Service; and the National Park Service. Focus is on coordinated grizzly bear management, including recovery planning.

Table 4.3. (*continued*)

Greater Yellowstone Interagency Brucellosis Committee (GYIBC). Chartered by the secretaries of the Departments of the Interior and Agriculture and the governors of Idaho, Montana, and Wyoming, the goal of the GYIBC is to protect and sustain the free-ranging elk and bison populations in the GYA and protect the public interest and economic viability of the livestock industry in the states of Idaho, Montana, and Wyoming. The committee meets three times a year.

Greater Yellowstone Winter Use Group. A group of representatives from forests and parks that work together on winter use issues. Current priority is monitoring impacts of winter use.

Greater Yellowstone-Teton Clean Cities Coalition. A regionally based group of public and private sector interests located in Yellowstone and Grand Teton National Parks and surrounding gateway communities in Idaho, Montana, and Wyoming. The primary goal is to address energy efficiency and the use of alternative, cleaner fuels.

Beartooth Highway Working Group. Representatives from Yellowstone National Park; Custer, Gallatin, and Shoshone National Forests; Montana and Wyoming transportation departments; and the Federal Highway Administration are working on jurisdictional issues and long-term maintenance and improvement plans for the Beartooth Highway. Key contact: Federal Highway Administration.

Source: Greater Yellowstone Coordinating Committee, *Greater Yellowstone Coordinating Committee Briefing Guide* (Billings, Mont.: Greater Yellowstone Coordinating Committee 2001).

it perceives to be best practices in the way it funds, staffs, and works with its committees. These choices tell us what GYCC's goals and priorities are, how well it is organized, and, at least in part, how effective it is.

First, the committee spends too little money on special projects annually. Its total allocation of $300,000 is less than three-tenths of 1 percent of the collective annual budget for all the management units in greater Yellowstone, which is $110,000,000 total. No doubt, some other money and in-kind support is also expended through regular agency budgets on projects within the general purview of the committee, although a precise accounting for actual costs is impossible to determine at present. Generally, the projects that get funded seem to have local rather than ecosystemwide or high-order importance, and they tend to focus on biological and technical data gathering rather than on the analysis, implementation, or appraisal functions. Many seem to lack overall strategic value in terms of advancing interagency and public coordination and integration or ecosystem/transboundary management, although many of them do meet the committee's strategic fit priorities. A few midrange-level projects are supported but very few higher-order, systemwide projects are promoted. The relatively equal distribution of funds among the management units each year indicates an emphasis on fair sharing of available revenues rather than allocation according to highest-priority needs within the ecosystem. This distribution fosters unit autonomy and reinforces existing fragmentation, foregoing opportunities to improve coordination and integration on a systemwide basis. Funding patterns and project support give the appearance that improving coordination means uniformity in projects, that if a weed survey project is funded on one unit, then other similar projects must be funded on other units. Given this type of thinking, there perhaps is even less ecosystem/transboundary thinking, deliberation, and work on the ground than appears to be the case. The overall impression given from the record is that limited resources are not being used strategically to best advantage.

It would be helpful to develop a scale to rate projects for their ecosystemwide significance, given the overall problems. One useful criterion might be whether the proposed project would significantly advance leaders' and workers' understanding and ability to best manage the whole ecosystem. Projects that produce a systemwide appraisal of key issues could be especially valuable. For example, at a high level, a fundamental appraisal of the GYCC's operating assumptions, organizational arrangements, and policy goals would be invaluable. Such projects would help demonstrate links among the units across management issues and at the same time give an overview of the region's trends, conditions, and projections.

Second, given its goals, the GYCC is grossly understaffed and is served by too few subcommittees. It could easily use another half dozen people to help members stay up-to-date with the many ordinary, governance, and constitutive challenges in the region and the options for their amelioration. It could use these people to prepare "net assessments," that is, overall assessments of how well the GYCC is doing, what is happening in the operating environment, and what options exist other than those GYCC members already know about. They could be the focal learning instrument for the organization. This would require considerable "boundary scanning" and analytic work on the part of a larger staff. The consequence would be that the GYCC could better link with the many working groups, committees, events, and processes in the region and be much more problem oriented and contextual in its work. Qualified staff could also do a lot of the higher-order, strategic, systems work that is sorely needed. Their appraisals of key issues could keep the GYCC well informed on the status of old and emerging problems. Selecting and retaining the special kind of staff needed would be critical; special conceptual and analytic knowledge and skills would be essential.

Working groups and subcommittees are valuable in solving problems and getting the work done, but the number that currently cooperate with the GYCC in some way is sparse, given existing and foreseeable challenges throughout greater Yellowstone. They also tend to be limited in their roles, responsibilities, and communication with the GYCC. Additional committees could focus on broader issues than most existing groups do, such as regional biodiversity; all threatened, endangered, candidate, sensitive, and rare species; and other systemwide themes, rather than just swans, bears, or weeds. They could do net assessments and make recommendations. They could be made up of agency and nonagency members. They could largely focus on higher-order work, coordination, and integration. What committees might be needed to serve the strategic needs of the GYCC has not been systematically investigated nor discussed by the committee, at least as documented in the record. Subcommittees are a major tool to solve problems, cooperate, and demonstrate best practices. Given the context of GYCC operations, there are many more opportunities for these parallel organizations to fit into the gaps between the existing agencies and other major organizations in the greater Yellowstone ecosystem to help achieve management and policy gains.

In sum, the projects funded, the staff, and the subcommittees could all be significantly strengthened to aid the GYCC's vital work. The situation is underorganized, given the goals of the GYCC and the region's growing challenges. A concerted, explicit discussion is needed to best use limited money, projects, staff, and committees to address high-order work.

Conclusions

The GYCC has several tools at its disposal to address management policy challenges to greater Yellowstone's sustainability. Chief among these are its problem-solving skills, its willingness and ability to cooperate, and its ability to demonstrate its vision to the world through best practices and to build a track record of such demonstrations. The GYCC presently cooperates with the managers of its constituent units within the structure and operations of the committee and the federal agency bureaucratic system. It has formal goals to cooperate at public meetings and with the public, but little genuine cooperation at this level takes place. Cooperation can also be shown via the way it participates in the decision process. It demonstrates its goals and commitment through its own deliberations and behavior at its own meetings, its track record of demonstrations as marked by its accomplishments and "milestones" over the years, including the "Aggregation" and "Vision" projects, and its funded projects, staff, and subcommittee work.

Cooperation is improving slowly in the GYCC over time. Demonstrations are improving, too, with a slow trend toward more focused and more strategic work. The arena is slowly organizing itself, in a small part because of the GYCC, for systemwide understanding and work. This increased cooperation is in part being forced on leaders by the public and experts as the arena and the context become more complex over time. There remains, however, considerable potential for the GYCC on its own initiative to upgrade both cooperation and demonstrations in significant ways. The GYCC can overcome tendencies to take a traditional, largely bureaucratic mode of conformity in its deliberations and work. It can actively seek greater scope, strength, and duration of cooperation and demonstrations in all its work, including those setting the standards as best practices.

The idea of working with and through subcommittees is a good one for the GYCC. The subcommittees form a loose "epistemic" community—a network of professionals who are competent and acting under a recognized authority—that includes other contacts and relations beyond just members of GYCC agencies. Because these professionals share a common outlook, practices, and energy with regard to a set of problems that they are competent to address, the epistemic communities that they compose are key to making future improvements.[44] The number and inclusiveness of epistemic communities can be significantly expanded.

To capitalize on positive trends, the GYCC can organize its meetings and its work and focus its attention on ways to upgrade cooperation and demonstrations substantially. It can organize its behavior to make more decisions from the standpoint of a comprehensive, common interest and explicitly strive to meet the highest standards of the decision process. It can seek to be fully problem oriented and contextual in its work. There is huge potential to upgrade both staff and committees for improved cooperation and to support demonstrations that promote real progress.

5 Overall Assessment— Leaders, Bureaucracy, and Context

An assessment of leaders and the bureaucracies and contexts within which they operate might help explain the patterns evident in the behavior and activities of the GYCC and other leaders in the region. Looking at these three factors—leaders, bureaucracy, and context—will point out limitations in the present modus operandi of leaders (under their ruling paradigms) and in the way that the situation (the arena and institutions) has been organized to detect and resolve problems. In this chapter I go to the heart of the greater Yellowstone debate about how we will use the region and its resources and who gets to decide. This value dynamic—how values are produced (shaped) and enjoyed (shared) and by which people and interests—is what leaders and citizens alike are struggling with as we determine greater Yellowstone's future. This kind of net assessment will refocus our attention on factors where real improvements are possible.

Leaders

One of the most conspicuous factors in the current greater Yellowstone management policy dynamic is the leaders themselves. We know a lot about leaders from the rich field of leadership studies and allied work in psychology, sociology, and politics.[1] In this section I look at leaders as individuals in general terms, no matter where they are located or in what sector they work, and elucidate some of the factors that influence what they do in their crucial roles in social and decision processes. I do not target anyone specific although the characteristics

I describe, in an effort to establish a profile of the kind of leaders we need, may resemble those of some individuals.

Leaders as Individuals

It is fitting to study leaders as individuals as we are seeking to understand what sorts of people rise to positions of authority. All policy concerns human behavior, and the personalities, beliefs, and values of leaders are critical. The perspectives people hold and act on, that is, their identities, expectations, and demands, very much matter in decision making.

Personality. Personality is the label we give to the principal traits an individual displays as a participating member of society.[2] One important distinction within personality is between character and aptitude. Aptitude refers to potential skills, whereas character is about desired value outcomes, that is, what is important to a person, such as respect, power, or affection, and the degree to which he or she has the assets of the total, healthy personality at his or her command. Some personalities are well integrated in terms of character and aptitude. In democracy it is preferable to have leaders with democratic rather than despotic character and appropriate aptitudes.[3]

One aspect of personality is subjectivity and subjective identifications. People see themselves both as individuals and as members of some aggregate or group they have chosen. Some people are loyal to a limited number of other people, ideas, or organizations, whereas others have loyalty to the larger community. Identification is the principal way to create and maintain group cohesion and the "us versus them" distinction that we all draw on for meaning. The "we" is at the very heart of all personal and political process. People's expectations about their interactions with society, which are based on their identifications with other individuals and with groups, can be inferred by observing the demands people make in the course of their usual behaviors. People's sentiments (feelings or emotions), which are often expressed in symbolic forms, are another manifestation of their subjectivities. Aligning leaders' and followers' subjectivities with a shared democratic ideal, common interest goals, and requisite aptitudes is essential for the emergence of effective leaders.

In addition, all individuals' personalities operate within *bounded rationality,* a term coined by Nobel Prize–winning economist, Herbert Simon, in reference to the cognitive limits of rationality. Typically, people are quite narrowly bounded.[4] *Leadership and Self-Deception: Getting Out of the Box,* a book by the Arbinger Institute, provides a good description of this phenomenon.[5] Bounded rationality has huge implications for leaders as they make decisions that affect society and resource management. There are always limits in decision making with respect to knowledge of alternative courses of action, their relative utility, and their consequences. We all operate inside "boxes," but some people have a way of recognizing the kind of box they are in and assessing how it affects their decisions and behavior. In some instances, people can transform their narrow, rigid boundaries to larger, more open, more inclusive perspectives. The larger and more open the box, the more these individuals can achieve. Most people assume that they are rational, although their decision making typically leaves out important considerations and causes poor outcomes. As the social sciences have come to understand, the "rational actor" model of individual behavior is a fiction, although most people still subscribe to it.

Bounded rationality comes about because people can only deal with the magnitude and complexity of the problems facing them by oversimplifying both the dimensions of the problems and possible alternatives. Rather than finding solutions that are optimal, people settle for solutions that merely satisfy, usually only in the short-term.[6] This leads to problem solving (and policy making) categorized as *incrementalism* or *disjunct incrementalism* by political scientist Charles Lindblom and by most of us as *muddling through*. Yet it is widely used by leaders, organizations, and citizens, not only in greater Yellowstone but everywhere.

One consequence of personality operating within a narrowly bounded rationality is that leaders cannot distinguish a "problem" from a "dilemma."[7] A problem can be solved within the frame of reference suggested by the nature of the problem, by past precedent for dealing with it, or by the use of existing action programs and policy. A dilemma, on the other hand, cannot be solved within the framework of assumptions contained in the way that it is conventionally represented. Dilemmas require reformulation to find resolution. People must abandon their habitual ways of problem solving ("get out of the box"), such as bureaucratic routines, and find new ways of understanding and acting. This calls for innovation, which is hard to do in bureaucratic organizations. In fact, organizations often force their leaders to set narrow boundaries for considering certain issues and permit only a limited set of possible cognitive styles. This does not bode well for inventing transboundary management policy for sustainability in greater Yellowstone any time soon. This task constitutes a dilemma on a governance or even constitutive level. It cannot be addressed "inside the box" as a series of ordinary problems in piecemeal fashion.

Beliefs. A major determinant of individual behavior is the belief system dominant in the society in which a person lives, as well as the beliefs of the many smaller communities to which that person belongs. These belief systems, or in anthropological terms, the *myth system,* bound people's perspectives, shape their personalities, and locate them in their world. In general, these systems make people's responses appropriate and consistent with the shared beliefs of their communities, according to Charles Taylor, a psychologist and historian.[8] Taylor notes that myths allow us to articulate meaningful expressions of self within a context by providing a crucial set of qualitative distinctions that form the basis of our judgments about what is worth doing and why. Myths are the backdrop of our lives, providing the frame of reference, whether we know it or not, for our political, rational, and moral judgments; our institutions; and the policies we pursue. Most people are unaware of the myths they hold, use, and live out daily. In greater Yellowstone, several conflicting myths are being played out in the "old West" versus the "new West" dynamic and in many other community distinctions that we make, including, for example, business, recreation, and environmentalism.[9]

Because myths are so central, the adjustment process to resolve clashes in myths in society requires governance and constitutive actions. Conflicts in myths are often played out symbolically. In fact, symbols are the outward manifestations of myths, and are vital to people's identities, expectations, and demands. How leaders understand this dynamic process and manage their own and other people's behavior, both substantively and symbolically, is a critical feature in managing resources in the West. Leaders who understand that this adjustment process can occur only with governance and constitutive change will be most successful in helping people, organizations, and institutions make the required adjustments toward sustainability.

One of the dominant myths held by leaders in greater Yellowstone is *scientific management* or *positivism.*[10] This school of thinking assumes there is an objective world that our senses can comprehend and measure realistically. It seeks laws to explain every phenomenon in the universe. It also assumes that positivistic science must precede decision making, and it calls for putting expert professionals in the driver's seat of decision making.[11] Bureaucracy is an ideal medium for perpetuating scientific management, as are the traditional natural resource professions (such as forestry, range management, fisheries, wildlife, and recreation), all of which had their origins in this epistemology. Thus, this myth is highly problematic in the resource management arena in the West and stands behind much of the poor decision making, weak leadership, and maladaptive bureaucratic behavior. Most leaders in greater Yellowstone had their training in positivistic science or at least have faith in its capacity to solve problems. This predisposes them to see matters technically and causes them to neglect contextual factors, namely, social and decision process.

Positivism has numerous inadequacies, which are apparent in its application in natural resource management policy.[12] Lung-Chu Chen, an international lawyer, points out that positivism fails to understand the notion of decision or choice in the social policy process.[13] It pays insufficient attention to the goals for which standard operating procedures and rules are devised, it fails to value the consequences of particular applications of the rules, and it fails to relate rules to the dynamic context of social and decision processes. Furthermore, it fails to grasp the normative ambiguity involved in "rules," fails to come to grips with the generality and complementarity involved in rules (especially constitutive ones), and fails to develop and employ adequate intellectual skills in real-world, contextual problem solving. Finally, it fails to mobilize leaders in the relevant intellectual skills to solve emerging problems. Yet, despite its limitations, positivism still prevails in the thinking and problem solving of leaders and staff of the various agencies responsible for managing natural resources in greater Yellowstone. It also dominates the philosophies and actions of most conservation and environmental groups, and it persists in much of the public.

Values. Delving even deeper into the beliefs and myths that determine behavior, we come to the values that are central to people's lives. The importance of values in management and decision making is illustrated in a recent study of the U.S. Forest Service.[14] Researchers Ingrid Martin and Toddi Steelman found that the agency's values conflict with its objectives. One body of research shows that the agency is being responsive, is shifting away from utilitarian uses of resources and the view of resources as instruments for the advancement of society, and is now focusing on conservation uses and preservation of public lands via more pragmatic means such as community-based initiatives. Another body of research, however, reveals that agency employees are resistant to change and highly constrained by bureaucratic rigidities; that the agency is still strongly entrenched in scientific, positivistic management, despite claims to the contrary; and that the old Forest Service approach, with its commitments to positivism, is still at the core of agency thinking. Martin and Steelman found these perspectives in Forest Service leader teams that they studied at three levels— Washington, D.C., office, forest level, and field offices. Leaders have

conflicting perspectives at all three levels of the Forest Service bureaucracy, but the positivistic approach predominates in all three.[15] Even if
they are so inclined, leaders seem almost powerless to change the values of their staffs or their bureaucratic cultures.

Other Determinants of Leadership Behavior

The behavior of leaders is determined in large measure by their
personalities, beliefs, and values, but other factors are also important,
such as leadership style, capacity for and attention to strategic tasks,
and the leadership discourse or narrative they create.

Leadership Styles. Leadership involves relations with followers and
material, emotional, and symbolic bonds used to get work done.[16] It is
the patterns in these relations that reveal the leadership style being
used. As mentioned in chapter 1, leadership styles vary along demo-
cratic/authoritarian, transformational/transactional, charismatic/non-
charismatic, and other dimensions. Harold Lasswell, one of the most
productive and creative social scientists of the last century, came up
with functional ways to understand leaders and managers, two roles
that may overlap. The principal job of leaders is to provide orientation,
although leaders and followers give and receive orientation from one
another. Every leader-follower relationship entails an exchange of one
kind or another in value terms (such as wealth for respect). Recently,
William Ascher and Barbara Hirschfelder-Ascher looked at leadership
styles and options using Lasswell's work. Their description and eluci-
dation showed that much of the contemporary literature on leadership
styles is just a restatement of Lasswell's ideas.[17] Lasswell classified
leaders functionally as "agitators," "administrators," and "theorists,"
three styles that are differentially democratic and responsible. Agitators are good at arousing emotional responses from followers. Administrators typically take on transactional and task-oriented leadership
roles, but they do not have the emotional need to transform organizations, nor do they possess the contextual knowledge that is a prerequi-
site to bringing about change. There is no strong motivation for
administrators to personalize professional interactions either. Theorists are the idea people. They come up with theories, ideas, and notions that may be picked up and used by agitators and administrators.
Certain kinds of leadership are needed in specific situations.[18] Different kinds of leaders tend to gravitate to different styles, and the challenge is to balance these approaches to fit the situation. Leaders in the
Yellowstone region, including those in the GYCC over the years, tend
to be successful administrators. There are strong disincentives in

bureaucracies for individuals to play the agitator role, and the theorist role does not have much opportunity for expression.

A central concern for organizations and for society is how to select and empower leaders who can inspire followers, empower them to be self-activating, and bring about needed change. It is important for organizations, as they choose new leaders, to make sure that all three functions (agitation, administration, and theorizing) are brought to bear on the needs of their organization. Finding, promoting, and sustaining leaders whose styles fit the context and demands for change in particular organizations requires deliberate recruitment and selection processes, training incentives, and other strategies.

Strategic Overview. Whatever their leadership style, leaders must have the capacity to formulate strategies and the patience and attention to implement them. Such strategic overview allows them to develop a comprehensive understanding of context and operations and attend to the higher-order tasks of management policy. As described in chapter 1, the skills of critical thinking, observation, management, and mastery of technical matters will help leaders accomplish these tasks. A comprehensive view makes possible the somewhat detached perspective needed to maintain a focus on the "big picture," clarify goals, and know what to do in practical terms to achieve them. The GYCC's formal, stated goals suggest that the group constituted itself to exercise this type of oversight capacity, to shape the future of the region more actively, and to help put society on a path toward sustainability.[19]

How leaders position themselves, given their goals relative to the context of their operations, determines whether they will succeed or not. Taking a strategic position is essential for high-level leaders. Frameworks exist to help leaders gain comprehensive, contextual understanding and strategic oversight and avoid missing crucial points or placing incorrect emphasis on their observations.[20] Striking a balance between comprehensiveness and selectivity is difficult. A view that is too general fails to identify and solve local, specific problems; one that is too narrow fails to see the influences of larger issues on small problems and the interconnectedness and cumulative impact of the many smaller, presumably isolated problems. A comprehensive view of greater Yellowstone would recognize that its inhabitants are part of a global community.[21] Without a comprehensive understanding, as Ronald Brunner notes, the context frequently "comes back to bite you."[22]

Methods exist to help leaders produce a realistic understanding of their operating environments. Useful frameworks can help them to focus both comprehensively and selectively on relevant features, on the process of decision making and the effective use of power, and on outcomes in terms of production and distribution of benefits and burdens that flow from their decisions. The GYCC does not have such a framework at its disposal at present, nor do most other leaders in the region. As a result, their present behavior exhibits four elements that social scientist James Scott says can produce weak or failed management policy.[23] First is the "administrative ordering" of society and nature through state-initiated simplifications. People, landscapes, and resources are understood in simplified, stereotypic ways. Second is a "high-modernist ideology," a variant of positivistic scientific management, "best conceived as a strong, muscle-bound version of the self-confidence about scientific and technical progress, efficiency, the growing satisfaction of human needs, the mastery of nature (including human nature), and above all, the regional design of social order commensurate with the scientific understanding of natural laws." Third is the abuse of authority by a more or less powerful state or agency that is willing to use its full might to achieve its self-interested policy. Fourth is a weak civic society that is unable to resist the state. The confluence of these four factors in greater Yellowstone is quite evident in many contexts, including, for example, grizzly bear management by the agencies. They explain, in part, why the agencies and the GYCC's members continue to be targets of criticism by pockets of public resistance to federal initiatives.[24]

Discourse. Another important factor determining the behavior of leaders is the quality of the discourse they create.[25] Discourse is a narrative or simply a story that people tell and retell to explain the world to themselves; it is often the principal means people use to communicate. The discourse used to create meaning largely structures how leaders think, talk, and act. Some narratives are open and offer ways for leaders to expand their understanding of themselves and the context and scope of their actions. Others are restrictive, bounded, and limit creativity, innovation, and learning for both leaders and followers. The GYCC members often communicate with one another about their work through the management stories they tell one another. For example, an official from the U.S. Fish and Wildlife Service narrated at one meeting an account of the management problems he faced— managing the distribution and abundance of elk, winter feeding, diseases, vegetation, biodiversity, predators, public relations (especially with hunters), and power relations between state and federal agencies.[26] Much of the context of the issues he sought to address, and even the outcomes he preferred, was implicit in his narrative. In fact, narratives are probably the preferred way for GYCC members to present challenging management situations to one another as opposed to systematic, analytic, problem-oriented, and contextual descriptions. They tend to draw on their subjective, implicit understanding of cases and contexts to create these discourses and struggle to cope by using a highly selective narrative that functions to reinforce popularly held perspectives within the group. This approach, although useful as a starting point for communication, ultimately blocks strategic discussion, comprehensive understanding, and focus on the high-order tasks they must undertake. Social scientists tell us that stories are told and have meaning within "interpretive communities."[27] Problem definition, therefore, "cannot be fully understood apart from the audiences to which it is directed and the [narrative] styles in which it is communicated," according to James Throgmorton.[28] Storytelling as rhetoric is the primary way in which most people, including the GYCC, communicate with each other. It privileges selected people and information, in this case the GYCC and its stories, at the expense of other people and stories, including the public, and nonagency professionals.[29]

Why do leaders and others operate this way? Discourse, policy narrative, or "practical storytelling" is one way to integrate politics, history, and biology into a package that can be easily understood and digested by others, according to John Forester, who studies policy narratives.[30] For many people, it is easier to marshal the various aspects of an issue into a story than to analyze them in a more abstract, theoretical way. Listeners must "consider and interpret carefully practitioners' stories so they can learn from them, so they can get tips, reminders and cues, as they come to see the messy situations of their own practice in potentially new ways," according to Forester.[31] There is no clear distinction between rationality and emotion in telling or listening to stories. Good stories can be captivating and instructive, particularly if the hearers have experienced critical parts of the discourse, but a captivating story that glosses over inconvenient issues can also be misleading. These stories may contain conflicting claims or tell of a search for practical answers. Telling and listening to policy stories can help people develop some capacity for political judgment.[32] Often though, this political judgment is based on an account of a problem and its context that is quite bounded. Listeners may not be fully cognizant of what they are doing as they interpret what is said and as they form judgments. In addition, according to Emory Roe, a policy researcher, such stories can be surprisingly resistant to being discredited simply by demonstrating that they are untrue in a particular case. They can only be invalidated by offering a better and more convincing story. In short, the GYCC's stories are about conflicting values and the fact that all values cannot be realized. They are about leaders' decision making or coping and resolution. The elk management story and others told by the GYCC reveal "critical pragmatism," risk taking, searching and striving, resistance and opportunism, and political drama. Listeners can see that they have been in similar situations, they can see how the storyteller wants to handle the problem, and they can compare it with their own views on the matter. The stories reveal in a concrete way the political and cultural drama being played out as the storytellers see it.

Implications

The GYCC is made up of individuals who largely share personality traits, beliefs, and values. This is not surprising given all the years of selection and self-selection that brought each of them to their present status. These shared features serve them well in working with each other and in their respective bureaucracies. They all appear as stable personalities and their identities, expectations, and demands fit a fairly narrow profile defined by the professional and bureaucratic situations they occupy. A more democratic component in the group's character, including a greater sense of responsiveness and responsibility to the publics they serve, would produce more common interest outcomes. Despite having risen through scientific management oriented organizations, the aptitude or skill potential of all the members is high within the constraints of their bureaucratic lives. Most are extroverted and easily approachable. They are all loyal to the GYCC as a team and, simultaneously, to their individual agencies. Their subjectivities are organized to support team operations, while staying independent and loyal to their units and agencies, although this split loyalty causes problems in decision making. Decisions are limited to those that favor both the team and a majority of the individual agencies.

GYCC members largely adhere to the beliefs and values of scientific management or a "techno-expert" philosophy of management policy and bureaucratic orthodoxy, although they are highly sensitive to the political nature of their work. Respect is a value they all seek. All seem comfortable wielding power. These beliefs and values impose constraints that limit the scope and depth of their exploration of management options. These beliefs and values also fragment their attention and reduce the chances of creating a strategic, comprehensive overview of management policy challenges across the region. They also preclude a focus on the high-order task of "future shaping" that also desperately needs attention.

The GYCC, as noted, is composed mainly of transactional administrators who have risen to top levels in their respective bureaucracies because they have managed their units well and have been loyal to their employing agencies' culture and to their superiors in the Departments of Interior and Agriculture. Their success is based on their record of "maintenance" management of their units, not on their achievements as "agitators or theorists" or as "transformational" leaders. In all cases, GYCC members learned their mode of leading in bureaucratic organizations with fixed rules, roles, and standard operating procedures and in complex, highly political contexts that for many of them were inimical to their core competencies. They seem to have learned to deal with the issues they faced, but they were not essentially comfortable with them. They and their bureaucracies adapted by adjusting their practices slowly over time using disjunct incrementalism, inching forward and adapting as pressure from outside forced change.

In this context, leaders seek unit autonomy, while coordinating to the degree that they can without violating the imperatives of maintaining discretion and control over their respective units.[33] Because of this, the GYCC typically defaults to the noncontroversial and unobtrusive role of information collection and takes no action in the face of complexity. It proceeds incrementally on relatively low-level issues, taking on few mid- to higher-order issues. Overall, it is conservative and cautious, which is a defensive strategy to protect its current beliefs and value holdings. To date, there has been too little genuine problem solving and cooperation and too few of the demonstrations needed to put the region and its institutions on a clear path toward sustainability.

The GYCC uses a discourse that favors the interests of agency leaders, technical experts, and their allies who are now in influential power positions in the region and elsewhere. It also meets the value demands of traditional interests (such as extractive industries or business interests) outside the agencies. Concurrence-seeking behavior among members dominates the group's present agency-dominated discourse, which excludes many considerations. This effectively prevents integration of all that is known about the people living in and influencing greater Yellowstone. For example, the present discourse limits or prevents a thorough appraisal of the agencies and their policy preferences. It reflects the self-interests of the central participants and marginalizes many people who otherwise share the goals of better government, better management of natural resources, and transitioning toward sustainability. Some of those people left outside the dominant discourse believe that the GYCC and its member agencies fail to attend to broad public interests. Granted, some marginalized people are less than clear about their goals and expectations, and their skills at expressing their demands may be less than ideal, but they do deserve a place at the table.[34]

In sum, the region's leaders find themselves in a complex and demanding situation. The GYCC's performance, however, has fallen short of its own publicly expressed goals and those of many attentive members of the public. Its current approach overestimates both the capabilities and benefits of current approaches (which are largely limited to positivistic science and bureaucratic managerialism), and it underestimates the likelihood that consensus can be achieved through other, more democratic means (such as genuine community-based open approaches). The present discourse reflects the magnitude of the stake that powerful economic and political interests have in maintaining the status quo. These interests are reflected in and reinforced by a weak problem definition that says that things are generally OK and that, if needed, change should be slow. Neither the temperaments of the leaders nor the reward-incentive systems of their bureaucracies currently support a faster transition toward sustainability. As a result, leaders are unlikely to develop an adequate response or solution to the growing threats to greater Yellowstone's environment and growing weaknesses in the governance and constitutive system in the foreseeable future.

Bureaucracy

The focus on leaders and leadership leads inevitably to a more detailed examination of another of the most conspicuous factors in policy dynamics: bureaucracy. What we know about organizational behavior can explain and predict the behavior of government as it seeks to coordinate management policy in greater Yellowstone. Resource management in the region is structured by many bureaucracies, both in and outside of government, that have built-in problem-solving approaches in the form of standard operating procedures (SOPs). Their structure and range of behaviors are deeply institutionalized. In this section I look at greater Yellowstone's bureaucracies, their problem-solving strategies, how they cope with uncertainty (that is, incompleteness of knowledge about problems and contexts), and how they structure the arena to perpetuate more bureaucratic control. I conclude by considering how well the bureaucratic system is working in the region. Understanding bureaucracies in greater Yellowstone provides insight into why its leaders and the policy system behave as they do.

Problem Solving

Bureaucracy is a problem-solving strategy, an organizational system invented and used by society to organize people and arenas to accomplish collective goals. Most organizations in greater Yellowstone use this means to structure their operations, especially those in government. In turn, bureaucratic imperatives determine what governments do and why.[35] A deeply conservative mode of organization, bureaucracy is a relatively easy way to coordinate large-scale activities in routine ways. The fixed rules, roles, and regulations of bureaucracies can be administered impersonally and from a distance and are effected through the SOPs and established programs embedded in these systems. Dependable bureaucratic performance relies on SOPs, which are the rules of thumb that standardize and coordinate the work of large numbers of people over large areas. They are typically simple so they can be easily learned and uniformly applied. They are essential for certain kinds of concerted action—routine tasks like maintaining a fleet of vehicles or road repair—although they can cause organizational behavior to become highly formalized, cumbersome, and maladjusted. In general, SOPs are relatively fixed and do not change easily or quickly, regardless of the demands of the challenges facing organizations. Rooted deeply in the reward incentive system of bureaucratic structures and cultures, they typically provide the operating norms, basic attitudes, and procedures of staff members. Resistance to change is common in bureaucracies where SOPs are strongly institutionalized.

Although bureaucracy may appear monolithic, it in fact often "consists of a conglomerate of semi-feudal, loosely allied organizations, each with a substantial life of its own."[36] Coordinating these entities can be difficult, especially in arenas like greater Yellowstone where authority and control are fragmented. Officials in bureaucracies typically reduce problems to standardized methods, programs, and decision processes, which unwittingly have blind spots and vulnerabilities. In bureaucracies new challenges are often crammed into old, familiar cubbyholes, whether they fit or not. The people who are allowed to participate in particular decisions are narrowly restricted to those considered competent to participate by agency leaders, their staffs, and allies. This approach continues the focus on technocratic, scientific management input and excludes many competent people and large bodies of knowledge, especially those from the social and integrative sciences, which could be very helpful to leaders in the transition to sustainability. Because of these limitations, what works in a bureaucratic model of a program does not necessarily work in real-life practice.[37] This causes delays and limits progress in understanding and addressing problems. David Mosse, an ethnographer involved in sustainable development, noted that "most agencies are bound to a managerial view of policy which makes them resolutely simplistic about (or ignorant of) the social and political life of their ideas."[38]

Natural resource analyst Steven Yaffee corroborated how comfortable it is for bureaucracies to behave in standardized ways: "Traditional ways of doing things have been tested by the realities of time, agency staff are accomplished at carrying out their tasks, and longstanding patterns of individual and organizational behavior create a predictable and energy-conserving reality for agency staffers and leaders alike. Organizations generally do what they do because they are administratively comfortable, politically and fiscally feasible, and legally allowable. Like all organisms, bureaucracies define, find, and protect a decision space— a niche—in which they are comfortable and can thrive."[39] For example, although agency leaders may choose varying topics for the agendas at GYCC meetings, most behavior at meetings is predetermined by well-established SOPs, patterns of loyalty, and modes of thought.[40] Meetings are limited to existing routines that make up the field of effective choices open to leaders. Meetings steer leaders toward small, insignificant, and anticlimactic decisions. This is a record that the GYCC shares with many other bureaucracies, including NGOs in the region.

In large organizations or in arenas with multiple organizations, as in greater Yellowstone, the sheer complexity of problems and relationships presents decision makers and managers with many challenges. Authority and control in many cases are not adequately apportioned to those who have primary responsibility. In turn, having responsibility for only part of an arena and a narrow set of problems causes individual and organizational parochialism, which leads to territoriality. Parochialism becomes institutionalized through selective attention to available information, hiring people skilled at only one kind of outlook and work, and job security mechanisms that keep people in place for many years. Other factors that promote parochialism are internal group pressures, distribution of rewards and incentives, and external pressures that reinforce the "us versus them" territorialism.[41] All this works against the development of the top-notch, versatile leadership needed for a transition to sustainability.

Another constraint on bureaucratic problem solving is that acceptable performance is often limited by unstated goals that are at cross purposes with the official, stated goals of the organization. Unofficial operating goals are seldom revealed, but they nonetheless influence behavior. Leaders and staff members who try to innovate face a forest of unofficial limitations holding them in the status quo. These constraints serve as a form of conflict resolution in that they keep everyone more or less on the same track. Conflict is also minimized by attending to acceptable goals sequentially. For example, when a problem arises, relevant subunits in the bureaucracy mobilize to address the problem within the constraints that they feel are most important. When the next problem arises, it is taken on by other subunits similarly responding to a different set of constraints. This pattern is repeated until it results in great complexity, as has been the case in the scores of species conservation problems in greater Yellowstone, including grizzly bear, wolf, and bison management. The GYCC and unit managers also tend to simplify problems through the analytic error of proceeding to a decision with inadequate or incomplete understanding of problems and their contexts, which is the most common error for government bureaucracy and its programs.

Most government programs are composed of routines, that is, rehearsed, practiced, coordinated responses to past problems. Each program is a response that the bureaucracy has available in its repertoire to deal with a problem (often cast in the language of "drama" with emotional investment, good and bad people in conflict, rising tension, and finally climax and resolution).[42] The search for programmatic responses exposes built-in biases, including value preferences, specialized skill training, and organizational SOPs, other routines, and program limitations. In most bureaucracies the number of programmatic options is quite limited. It would be almost impossible to invent a new program for each new problem, especially governance and constitutive ones. The rule of thumb is that the more complex the problem and the action required, the more important it is to depend on established programs. The complexity of transboundary management reveals the difficulty leaders might have in deciding which programs and SOPs should be mobilized or changed. The chief risk for leaders and bureaucracies is uncertainty, and the preferred way to deal with it is to routinize responses and agency behavior, especially if the arena is a contested or negotiated one. In established arenas, like the one created and maintained by the agencies and the GYCC in greater Yellowstone, interactions can be stabilized and "club relations" can provide a comfortable setting in which to operate. Standardized programs are bureaucratic solutions to problems that are too complex for any single person to understand; they help individuals cope with their own bounded rationality. The unstated goal here is not to solve problems but to cope with uncertainty and avoid personal risk.

Existing bureaucratic routines, SOPs, and existing programs are the stopping point in the search for alternative courses of action. It is nearly impossible for bureaucracies to go beyond these in their search for responses. Innovation is thus blocked, and organizational learning is extremely limited since it would be too threatening to routines, SOPs, programs, and to leaders and allies.[43] Focusing on recognizable cases, one at a time, and seeking at the same time to control and adapt them to fit into existing SOPs is the basic bureaucratic response. As a result, the bureaucratic status quo inevitably stays in place.

Coping with Incompleteness

Another characteristic of bureaucracies that is evident in the GYCC is their astonishing ability to cope with uncertainty and incomplete information. All challenges confronting leaders contain uncertainty, which stems from "incompleteness," among other things. Incompleteness in the daily working life of leaders is about the unending stream of new information, problems that must be solved, obstacles overcome, goals achieved, brush fires put out, constraints overcome, objectives attained, missions met, and crises weathered, such that management is a life of endless interruptions and unfinished work.[44] As a result, almost all of the work undertaken by leaders is left incomplete, which makes addressing problems risky.

Emery Roe, a policy researcher interested in sustainable resource development, devised a framework to assess "incompleteness" in resource management policy and to classify leaders' responses to it.[45] Incompleteness is not the same thing as complexity or uncertainty. Complexity exists when there are many varied and interrelated elements involved, and uncertainty refers to unclear or poorly understood causal processes. But incompleteness is about interrupted, postponed, or unfinished aspects of the work at hand, all three of which figure prominently in the life of leaders in greater Yellowstone, especially the GYCC.

For example, greater Yellowstone is ecologically complex in part because its internal operational principles are uncertain, that is, their casual relationships cannot be fully explicated. This makes decision making about management issues incomplete. To many people, this fact calls for more positivistic scientific research, with the expectation that someday we might understand the causes of all ecological processes and thus how to manage optimally. This viewpoint assumes that complexity leads to uncertainty and that uncertainty can be reduced only through positivistic scientific study, which then informs management policy improvements. This linear sequence—that science must precede management and policy—is the scientific manager's understanding of the relation between science and policy. But, in fact, most management and policy results from practice-based knowledge, not positivistic study.[46]

Given that incompleteness is pervasive and cannot be eliminated, how do leaders stabilize their assumptions for decision making? As Roe and other authors note, leaders draw on less formal "policy narratives" or discourses to organize their understanding of events and processes.[47] They draw on theories that do not deny incompleteness but nevertheless allow them to take action that somehow accommodates it. Some policy narratives serve as shorthand for shared assumptions about management policy, leadership styles, and ways of coping with incompleteness. On one hand, different narratives deal with these features more or less realistically. A policy narrative that fully appreciates these features without paralyzing leaders to inaction is a valuable one. On the other hand, a narrative that neglects these features can lead to management disaster.[48] It is these narratives that leaders, unit managers, and the GYCC invent and use to deal with incompleteness in the greater Yellowstone ecosystem. In GYCC the narrative is one about scientific management, bureaucratic discretion, and decision authority. The narrative in use strongly reflects the policy discourse that leaders employ. Ideally, discourses should genuinely fit the nature of the challenges—ordinary, governance, and constitutive—that leaders face. What we see in play in greater Yellowstone is a mismatch between leaders' discourse and real-world challenges. The old narrative retains a powerful grip on leaders while new problems proliferate.

Roe devised a "threshold-based resource management framework" to help leaders cope with incompleteness.[49] For example, when human population levels reach certain thresholds or when resource uses reach certain points, then a different kind of leader and management regime may be needed. Put another way, each management regime is an all-encompassing narrative—a metanarrative—which includes a certain kind of discourse and a model of learning. The theories, inherent models of learning, and metanarratives that are useful at one level may be nearly useless, or even counterproductive, at another level. Holding on to an ineffective narrative dooms any management regime to failure.

Roe describes four possible leader/management regimes. The first approach is self-sustaining management, wherein the system largely takes care of itself. A good example of self-sustaining management is wilderness areas, which have few people, no extraction of resources, few multiple resource uses, few causal models of how the system works, and high ecosystem health. Situations like this, where management does not play as crucial a role, are rare in any greater Yellowstone ecosystem.

The second approach is adaptive management, which may survive with a trial-and-error or experimental approach. Examples of this type of management exist in most national parks and forests. Adaptive management is typically done incrementally, using positivistic, science-based models, either passive or active.[50] Seldom is adaptive management done formally. It usually takes place around specific issues, as it does in greater Yellowstone around the restoration or preservation of selected ecosystem features, such as preserving grizzly bears, eliminating invasive species, or restoring natural fire regimes.

The third approach is case-by-case management, which occurs in zones of conflict where people, resources, and environment increasingly compete. This situation is characterized by rapid population growth, increased resource extraction, and rising pressures for more user amenities like recreation—exactly the situation in greater Yellowstone.[51] This is overwhelmingly the dominant approach used by the GYCC and other leaders in greater Yellowstone.

The fourth approach is high-reliability management for intensely used ecosystems (such as intense agriculture, nuclear power plants), requiring error-free management, highly skilled leaders, and highly reliable organizations.[52] Such systems, according to Roe, require high technical competence, high performance and oversight, a constant search for improvement, highly complex activities, high pressures, incentives, and shared expectations for reliability, hazard-driven flexibility to ensure safety, and a culture of reliability (which cannot be replaced). This kind of management is a special case and has very limited applicability to greater Yellowstone.

Each regime or framework is really a metanarrative or discourse that contains, first, simplifying assumptions that permit leaders and managers to proceed in the face of incompleteness, and, second, a way to deal with their own bounded rationality. As we have seen in greater Yellowstone, an overreliance on the case-by-case approach and underuse of strategic thought and action have blocked exploration and possible adoption of approaches beyond the bureaucratic ones currently in place. This compromises the overall quality of the ecosystem, puts the agencies in a bad light with the public, and renders effective, first-class management nearly impossible. It keeps leaders trapped in a system of thought and action that holds tightly to the status quo.

Organizing

If the GYCC is serious about transitioning toward sustainability, it must consider how the arena in which it operates is organized. The committee must consider whether the greater Yellowstone arena is sufficiently organized at present to facilitate a smooth and relatively rapid transition toward sustainability, and if not, how it could be better organized for the invention, diffusion, and adoption of transboundary management. The current arena, built up over decades in a piecemeal fashion, is well configured to benefit the status quo, bureaucracy, and special interests.

Arenas. Leaders have the capacity to organize the arena to advantage. The GYCC's record (of meeting agendas and minutes and discussions) does not explicitly consider the arena concept. There is no recognition in their meetings about whether their arena—greater Yellowstone as a whole, its many smaller composite arenas, or its intersections with other arenas in the region, country, and beyond—is adequately organized or what, if anything, could be done to organize it better, given their goals.

Considerable thought has been put into conceptualizing arenas and their consequences on policy behavior.[53] An arena, again, exists whenever people's interactions affect each other, whether they are aware of the impacts or not. Arenas can be understood as centralized or decentralized, continuous or short-lived, focusing on specialized topics or general interests, organized or unorganized, and open or closed to broad participation.[54] The ideal is for them to remain sufficiently flexible to balance all these dimensions, which allows for the broadest community goals to be realized in practice. Little progress is possible toward sustainability without an appropriately organized arena. The present pattern of interaction in greater Yellowstone is not well balanced across competing values and claims, largely because of leaders' inattention to this vital matter.

Inclusivity. When people's interests converge, communications among them increase; an association, committee, or subcommittee springs up; and this is how an arena becomes organized. We can look at patterns of participation, inclusivity, representation, and responsibility to track these changes.[55] The greater Yellowstone arena is, overall, slowly organizing for many reasons, in part because of a few people in the agencies and the GYCC, but largely because nongovernmental interests are organizing to advance their own agendas (including the Greater Yellowstone Coalition, Yellowstone Business Partnership, and many others). Few of these changes, however, effectively challenge the dominant discourse or the status quo governance and constitutive formulas now in place throughout the region. In fact, any changes face powerful status quo and bureaucratic forces that would restrict change.

Although participation in organizations and associations is growing, at present there is little opportunity for individuals or organizations to participate in GYCC meetings and deliberations in meaningful ways. Individuals can and do participate in diverse ways in natural resource management issues with agency staffs and among themselves (for example, via media, e-mail, campaigns), but their participation is often of limited utility and is typically formulaic. The core demand from the public and other specialized sectors (such as academia) for respect and for inclusion typically goes unmet by unresponsive agencies and elected officials. In fact, no forum presently exists anywhere in greater Yellowstone to talk about the shortcomings of the present arena and its governance and constitutive processes.

All individuals who either affect or are affected by the decision-making process should share in the exercise of power and should also be subject to that power. In other words, arenas should be inclusive. In GYCC meetings, for example, in the portion of the meetings that the public is permitted to attend, the audience should not be excluded from deliberations. At present, the expectation is that the audience will be quiet, passive listeners, except when given permission and time to talk. This is hardly a model of inclusiveness.

Inclusivity involves two other considerations—representativeness and responsibility. The first involves a determination of who should be represented in deliberations throughout the ecosystemwide arena, in the GYCC's arena, and in other specialized arenas. Basically, all affected groups should be represented in the decision-making arena, especially those who are significantly affected by decisions or who have something significant to add. Plurality should be encouraged. The second consideration, responsibility, is measured by community standards that must guide all interpersonal interactions. Individuals who are not willing or capable of meeting these basic standards of responsibility to their fellow citizens should be denied participation. Maximum equality should be sought. The GYCC's meetings, generally speaking, are highly selective and do not permit the kind of representative, responsible participation that is ideal. Standards for representation and responsibility—indeed, the quality of social interactions generally—at GYCC meetings and throughout most of greater Yellowstone's arenas fall short of the goals that the GYCC says it seeks.

Standards. It is important for leaders to develop standards by which to measure how well an arena is organized. First is the centralization of decision making.[56] Since neither a fully centralized or decentralized arena is ideal, the two extremes must be a balanced in order to suit the context.[57] In greater Yellowstone today, generally speaking, the arena is overly centralized.

Second, the effectiveness of an arena's decision making and the types of problems that can be effectively addressed depend on the arena's longevity. For example, higher-order problems related to transitioning toward sustainability through transboundary management policy require long-term, continuous arenas. Short-term problems, such as allocating money for annual weed management, can be addressed by a short-term arena. Specialized arenas should be designed around particular issues in order to match the nature of the problem. Greater Yellowstone needs more properly organized, continuous arenas, especially those that would benefit from reliable forums of participant interaction and genuine joint problem solving.

Third, the benefits of organizing an arena for effective decision making depend on how tightly the arena is structured. More structured arenas permit decisions to be made efficiently, whereas more loosely organized arenas allow more open, flexible, and creative policy deliberations. Each arena should be structured for an optimum balance between tight and loose organization. For example, public education requires enough organization to standardize the educational content and ensure quality education. But it is also helpful to be appropriately flexible and unstructured to allow for adaptation by small groups for individual interactions and differences in learning styles. In greater Yellowstone there are some arenas that are overorganized and some that are underorganized; with more explicit oversight they could be appropriately organized depending on the issues and contexts.

Fourth, arenas must be balanced between specialists and generalists so that optimal decision making occurs. A key variable is whether the arena is dominated by experts or whether it is more general in composition. In a specialized arena, expert-driven decisions may be best. For example, an expert might be asked for a technical opinion on the likelihood of meeting a goal by certain management steps when the goal is agreed upon by the community. But in a broader arena with more inclusive participation, experts should not drive decision making. For instance, in a nonspecialized arena, where there may not yet be consensus on goals, broad participation is important in order to achieve an outcome that will most likely serve the community's interest. Elk management in western Wyoming is an example where neither the generalized nor the specialized arenas are working very well. In this case, although the goal had not yet been agreed upon, government nevertheless set up a specialized arena that it could dominate. Furthermore, the federal and state governments are in ideological competition over the nature of the arena, the decision-making process, and the preferred outcomes. As a result, the overlapping arenas are ineffective, and emerging policies have little public consensus and are seen by many as illegitimate and not in the common interest.

Fifth, arenas must be organized in terms of greater or lesser inclusivity. Opening arenas so that any particular participant can be part of them is basic to democracy and open public deliberation. Access to arenas can be made easy by creating a climate that welcomes participation. It is also possible to make access difficult, by, for instance, scheduling meetings far away from the people most interested so that they cannot travel to attend. However, geography remains a very real challenge because greater Yellowstone is such a large area. Ideally, arenas should be open and available to everyone who has a legitimate interest in the outcome or a role in decision making. Participation should be compulsory for people who are critically affected by the outcome.[58] Technical, specialized arenas should be used only to the extent needed. Arenas that consider decisions affecting the broader community should be open. In greater Yellowstone, there are too many arenas to which access is limited, participation restricted, and competency delegitimized by government agencies. Restrictions like these abound at a time when broad participation on management policy is needed.

Implications

Bureaucratic dominance in the greater Yellowstone arena has huge consequences for natural resource management policy, leaders, the GYCC, and the public interest. Bureaucracy brings with it deeply institutionalized problem-solving strategies, embedded modes of coping with problems, and fixed ways of interacting in diverse arenas involving the public and experts.

First, the federal bureaucracies and, less so, the state agencies dominate policy making for managing natural resources. They use SOPs and programs in preset, rigid ways to address problems. The political imperatives of bureaucracy, including the drive to retain autonomy and defend their turf, are hugely important in all interactions. These constrain the discourse and what is considered permissible to discuss and do. Bureaucracy has important consequences for how the arena is set up and maintained, including who is recognized or designated competent by bureaucrats to participate in which arenas and discuss what subjects. Overall, bureaucratic dominance maintains the status quo and limits flexibility, creativity, and broad participation in both general and specialized arenas. As a result, the common interest suffers.

Second, leaders are not coping adequately with the incompleteness of issues and the press of heavy workloads. Their undue dependence on bureaucracy is one principal reason. The management demands imposed by leaders' home units limit attention to the work of the GYCC, interunit coordination, and the operating context. There is only one dedicated staff member to aid the GYCC in addressing these broader issues. To cope, leaders manage on a case-by-case basis. There is little effort to be comprehensive or strategic or to attend to higher-order leadership tasks, resulting in a piecemeal, incremental response. Overall, this coping style produces less than effective management policy.

Third, the overall general arena of greater Yellowstone, as well as the specialized arena of the GYCC and smaller component arenas, are organized suboptimally to facilitate a smooth transition to sustainability via transboundary management. Leaders typically arrange and participate in these diverse arenas in ways that advantage and privilege themselves and their home units. There is some token discussion of broader issues as a way of preempting other arenas from taking them on in a substantive way. There is almost no explicit attention to how best to organize the GYCC or any other arena in ways that could facilitate pursuit of the public interest.

Context

The third major factor in the current policy dynamic around managing greater Yellowstone is context. Functionally speaking, context includes the current pattern of social interactions, which is the principal determinant of the scope, strength, and duration of cooperation that is possible within an arena. The global, national, and regional contexts were introduced in chapter 1 in conventional terms. In this section I look at the constitutive policy process in greater Yellowstone and the nation in functional terms. The constitutive context structures the present "policy regime," shows chronic problems, and at the same time offers opportunities for adaptation, change, and sustainability. It behooves leaders and citizens to be fully aware of context and the constitutive process.

Understanding Context

Scholar Gregory Bateson said that leaders, scholars, and the public do not sufficiently appreciate the significance of context.[59] This is certainly true for most people working in greater Yellowstone today. It is easy to misconstrue the context and its importance and hard to know what to do about a changing context full of problems and opportunities.[60] The greater Yellowstone region is undergoing mushrooming demands from a growing population with diverse and seemingly incompatible interests. Plant and animal communities are also undergoing both natural and human-caused changes. Leaders are struggling with this dynamic context, all the while trying to work through the cumbersome bureaucracies that have evolved in the GYE.

The context affects how or whether the GYCC—as a body, as unit managers, and as individuals—seeks out cooperators and deals with crises in management and policy that arise. Among key contextual factors are "incidents," including the congressional hearings of 1985–86 and the 1988 fires, as well as various legal challenges and long-term trends, such as the burgeoning demand for oil and gas. My observations and the written record show that the GYCC does not carry out systematic, explicit contextual mapping for its own uses, nor does it communicate internally about these matters in any explicit or systematic way that is evident. As a result, GYCC members do not share a common contextual framework, much less a shared view of constitutive process, at least not to the extent needed for effective leadership.

Contextual mapping requires determining who is involved with problematic situations and what are their perspectives, their values and strategies, and the outcomes they seek.[61] For example, what is the invasive weed problem, not only in the biological or technical sense, but also in terms of the social and decision dynamics that affect plant and seed dispersal and how to address these problems? It is difficult enough to map the context of an ordinary problem realistically, but it is much harder to map the context of a governance or constitutive problem in real-world ways. Yet, as policy researcher Elaine Castle has pointed out, contextual information "is the primary resource, in the sense that without it we lack even the most elementary tools with which to solve present problems and build a better future."[62] Having contextual knowledge is vital to problem solving, cooperation, and leadership.

More specifically, mapping context means carrying out descriptive and analytic exercises that reveal the relation between the parts and the whole of any problem under consideration.[63] The resulting map is

available for all people to see, talk about, and use in their deliberations, choices, and implementation. Without a shared, realistic contextual map, leaders and the public will be operating with incomplete
and different maps of what is happening and why. They risk working
at cross purposes. The underattention to contextual mapping by leaders, including the GYCC, severely limits their effectiveness, thus
retarding civil discourse, cooperation, and problem recognition and
resolution at important scales.

Constitutive Context

Almost all of the many ordinary problems in greater Yellowstone,
as well as the governance problems, stem from the very makeup and
dynamic nature of our culture, that is, the social and decision processes that give us its distinctive character, or in other words, the constitutive process. Because demands on greater Yellowstone's natural
resources will grow with time, we must understand the existing *resource regime,* that is, the system of resource use, how it operates, and
what future prospects look like. This regime is a direct manifestation
of the constitutive process over history. It is clear that the constitutive
process of decision making is by far the most important social process
underway in greater Yellowstone. It is also the most unformed and un-
stable. It is important because understandings built up over the last
century represent the most significant, persuasive strategies for inclusive shared use of greater Yellowstone's resources. We must also appreciate the constitutive process at play today, the chronic problems
that are emerging and becoming clearer because of the way the process functions and the growing demands by the public for change, new
leadership, and for new management policy.[64] What follows can serve
as a starting point for an enlarged, more inclusive discussion and a
realistic, shared mapping of the context.

Resource Regime. The patterns of use of greater Yellowstone's natural resources have changed over time. Today these patterns exist in the
form of laws and other norms, the structure and behavior of government and private organizations, and the actions of citizens. We all can
see many of these features. In order to describe the regime in more
functional terms, however, we need to understand that the present balance of resource use reflects conflicting claims about who should control resources for inclusive and exclusive uses. For example, public
lands are shared by many users, but recreational use of those lands is
available to all, whereas livestock grazing on those lands benefits only
those ranchers who graze animals there. Inclusive use means that

access is open to everyone; exclusive use means that use is reserved for special interests. In short, the constitutive process allocates these kinds of uses of resources. A pattern of claims and counterclaims is evident in greater Yellowstone today about access to natural resources. It is underway in the contentious public debates that we see in the media, the courts, and on the street today.

Greater Yellowstone is clearly a contested landscape. Issues about claims and access are at the very heart of our culture and the principles by which it functions. The constitutive process, which resolves such issues through social and decision-making processes, has over time given us our present suite of laws, prescriptions, and customary practices for the use of those resources. The constitutive rules that guide use of greater Yellowstone are not a static code but a continuous flow of interactive decisions. The GYCC and its unit managers are just some of the many decision-making bodies that establish and maintain this continuous process of decision making about resource use. This constitutive process shifts or adjusts over time in varying degrees in response to the claims and counterclaims being made at the time by varying interest groups. In other words, the constitutive process is responsive to the context. This process has allocated some resources to the exclusive use of some interests (such as business concessions in Yellowstone National park, grazing leases on Forest Service lands, and oil and gas leases on other public lands) to the exclusion of other people and interests. Other resources are inclusive, shareable resources in the present regime, including air, water, scenic views, recreation, and wildlife, which are open to and provide benefits to the larger community (that is, the citizens of America and the world). In the Yellowstone arena, exclusive and inclusive uses exist in a complex mix. In general terms, the conflict that we see everywhere around us today is between proponents of local control, who tend to promote exclusive use, and proponents of broader concerns, who tend to promote inclusive use.[65]

One of the most important tasks of the present constitutive process is to achieve an appropriate balance between inclusive and exclusive interests. This means weighing competing claims and uses in a great many ordinary and governance processes.[66] In greater Yellowstone this battle has often been resolved historically in favor of shareable resources under inclusive use, as evidenced by the large amount of public land in the region that resulted from past constitutive decisions. Yet the struggles continue. Many decisions, particularly those concerning wildlife, have been left to the exclusive mandate of the states. With regard to nonrenewable resources, such as biodiversity (plant and animal communities and entire ecosystems), the constitutive process has not been as responsive as needed with the exception of the Endangered Species Act, which has proved to be not as sensitive nor as effective as anticipated originally. No national inclusive policy yet exists specifically for biodiversity.

Resource use patterns have been fairly stable over recent decades, compared to the longer term of the last 150 years in greater Yellowstone. During the earlier years of that span, the resource regime changed as dynamically as our rapidly growing country. These changes have been in terms of investments and trade (e.g., market, ownership, land uses) and also in the means that were developed to protect natural resources (including establishment of the federal land and wildlife management agencies, the Endangered Species Act, and other legislation). Intervals of stability and instability have existed.[67] The early regime was locally based on the fur trade, livestock grazing, logging, mining, settlement, and tourism. Until the establishment of Yellowstone National Park, which was followed by the forest reservation system and eventually the U.S. Forest Service, the use of regional resources went unregulated, formally at least. The major change has been away from extractive resource uses and toward an economy dominated by tourism, retirees, and independent incomes.[68] These patterns are partly controlled by the market but are also influenced with varying success by the power of elites, a clear example of which is the Rockefeller family's use of power to shape regional arrangements in ways that reflected both their private and public values. In the 1940s the Rockefellers purchased land on the floor of Jackson Hole and gave it to the U.S. government; this land eventually became an important part of Grand Teton National Park.

Over time the regime has always allocated "competencies," decision-making power, and resource use to meet the needs of elites in society. As used here, competency means that people, groups, and organizations are recognized by the elites (within and outside the agencies) as capable of participating in resource use patterns in ways consistent with the demands of those decision-making elites. For example, Yellowstone National Park was conceived by Eastern elites, who thereafter brought new competencies into the region to manage its resources (e.g., federal agencies). Added to this trend over the following decades were the U.S. Forest Service and later the U.S. National Park Service as they were created. Academia has given rise to other potential participants with new competencies. There have been many other changes more recently. For example, greater Yellowstone is served by scores of NGOs presently. Today resources are shared through social regulation by the federal and state governments and other institutions in society. The proliferation of potential participants means that determining who is competent to participate in the regime's deliberations, and in the constitutive process itself, is a matter of great contention.

In looking to the future, shareable resources such as greater Yellowstone must be managed in ways that require greater levels of cooperation than ever before in the region's and the nation's history. Perhaps appreciation of this imperative is why the many leaders in government, nongovernmental, and business groups have striven so hard to work together recently. In a real sense, the present situation requires cooperation among regional communities, state and federal governments, and in some cases the world community (given global markets and global events such as climate change).[69] Sharing of resources as it occurs today is what we might call "a working specification of the common interest." Yet there are changes underway now and in the foreseeable future that call into question the viability of this existing approximation of the common interest. In response to the changing context, the constitutive process will evolve new rules and norms, thus creating a new "working approximation."

The federal agencies set up the GYCC ostensibly to transition the region's policy regime for managing resources into something more appropriate to the changing context. This transition is typically discussed in terms of both better coordination and integration and also new transboundary management. The committee, ideally perhaps, hopes to make the needed adjustments through new management policy. The shared common interest in greater Yellowstone is to provide more values to the larger community. If the GYCC operates successfully on a constitutive (and governance) level, it can lead us toward this new, more sustainable management policy. Shifting the present regime to a new pattern that permits continuation of shareable resource uses without destroying them is the principal challenge before the GYCC and other leaders, both in and out of government. The outcomes and effects will depend not only on what the GYCC does but also on the national and international constitutive processes.

Historically, the unregulated pattern of resource use policy in Yellowstone was guided by the view that its resources were inexhaustible. We know very well today that these resources are quite finite. The view of the inexhaustible abundance of resources is now incompatible with present facts and theories about resources and their sustainability, including the relation of resources to technology and our sociopolitical system of democracy.[70] Nevertheless, many people at local and state levels contend that greater Yellowstone's resources should be more open to exclusive use by local business and special interests than they presently are (for example, the privatization movement, Yellowstone Business Council, and the Political Economy Research Center in Bozeman, Montana).[71]

Whether greater Yellowstone's resources should be used inclusively by all people or exclusively by business elites and others is at the heart of the present public policy debate about the future. Again, it comes down to the questions of how we will use resources and who will have the power to decide. It is about how governance and constitutive processes should operate and to whose advantage. How to accommodate competing demands is the present governance and constitutive challenge. Whether this can be resolved through explicit agreements among the contending interests or whether coercion (power, regulation) is required remains to be seen. Probably the end result will be a mix of the two. Regardless, top-notch leadership is needed to guide society through these difficult times. The preferred new regime for greater Yellowstone, one based on cooperation and persuasion, will help bring about more productive, more equitable, and more sustainable social and economic conditions than exist today.

Most management policy today deals with the allocation of the greater Yellowstone ecosystem's internal resources. However, many of the resources flow in and out of the region. For example, the quality of the air and rain that flows into the region from the west and the water and soil that flow out (erosion) connect greater Yellowstone with the rest of the world. So do animal migrations and plant dispersals. Historic agreements about internal resources have been the best means of sharing resources. The GYCC's job is to extend and improve on these constitutive agreements to meet the needs of a wider world. Even though many of these agreements remain controversial, including those regarding oil and gas, livestock grazing, endangered species, and large carnivores, building on current agreements will give us the best chance to meet minimum public order goals of our society in using resources.

Chronic Problems. In the United States we have escaped many of the terrible problems that plague the rest of the world, so far at least, but in the GYE we have a stake in assuring that our system of public order (produced by the constitutive process) does not fall short of people's hopes, as they have in so many other parts of the world. What are the chronic problems? It is clear today from the growing intelligence about the world's environment overall that there are many serious problems with regard to human rights and the sustainability of the human enterprise. Current structures, processes, and practices have produced inequitable distribution of resources, goods, and services. In some regions, there are huge discrepancies between people's aspirations in terms of human dignity, prosperity, and freedom and their actualities, which too often include hunger, ignorance, human rights abuses, war, displacement, pollution, and depletion of resources. The breakdown of agreements, as evidenced by poliferation of conflicts— from lawsuits to wars—indicates growing uncertainty about the stability of the present system. World population growth further compounds the challenges, putting immense pressure on scarce resources.[72]

There is controversy about the status of natural resources everywhere.[73] Some people believe that scarce resources are not a problem, that technology will open up new possibilities, and the so-called limits to growth will not limit the human enterprise. Many in greater Yellowstone hold this view. Others take the opposing view that overreliance on science-based technology and the unintended consequences of technology have caused many of our present problems and that it is unlikely that technology can solve the problems it created, much less the long-term ones. They argue that governance and constitutive problems cannot be solved by science and technology.

A growing number of people are convinced that there are in fact limits to the growth of the human enterprise, especially in the numbers the earth can support.[74] They see that the present system of public order is powerfully organized toward short-term, exclusive, special interests that favor wealth and power elites. It is this system that produced government agencies to ensure predictable flows of resources into society. This system does not rely on rationality for the allocation of resources over the long term. It will be hard to change the dominant governance and constitutive process to move toward sustainability. Some people doubt that human nature will permit the level of cooperation needed to transition toward sustainability.[75] Like all organisms, people tend to operate in ways that they perceive will leave them better off in the short term, unless they have a clear sense of dire long-term consequences. The present situation will continue to stress humanity's capacities, as reflected in our sociopolitical systems and institutions. These are severely overstressed in Haiti, for instance, as well as many parts of Africa and other parts of the developing and developed world, and there are indications that even regions like greater Yellowstone will be affected as well.[76]

Demands for Change. In greater Yellowstone, and throughout the world, people are increasingly dissatisfied with the outcomes of the resource allocation process. Governance and constitutive decision processes lack the clarity and adaptability to adjust these processes to real-world contextual demands. In greater Yellowstone, people are organizing themselves through NGOs and in other ways to make their demands heard. This complicates the job of leaders, but it also informs it with urgency and energy.

These new demands are being made now primarily because the public is becoming more conscious of what is happening and what is at stake. People are more aware of the unequal distribution of resources as reflected in different standards of living and privileges (for example, influence and wealth) of some segments of society. People are more aware, too, of the discrepancy between the standards that should be promoted to serve common interests and those that are actually carried out in pursuit of special interests. This consciousness has resulted from an explosion in new information and communication from universities, NGOs, and others. These new demands directly challenge older patterns of resource use, the old resource regime with which the agencies have long been aligned. In greater Yellowstone it is the exclusive resources that are the focus of most claims, making for new arrangements and prescriptions to make these resources more inclusive. The judicial courts as well as the court of public opinion are being used increasingly to make or defend claims. These are forcing a new direction in natural resource management policy.

Implications

Looking at greater Yellowstone as a region and as an idea, and focusing on its resources, its people, and their interests, we see that past constitutive processes that allocated resources are increasingly failing today. To fix the problems, we must learn to judge the outcomes of decision making and leaders' behavior in terms of the aggregate, long-term, common interest. This requires us to understand the context fully and realistically, which can lead to several advantages. First, it seems obvious that if we are truly committed to human dignity in a world where resources are limited and human demands are unlimited, we must allocate resources more rationally over the long term. Second, we must formulate a more appropriate process of decision making—both governance and constitutive. We need better common interest policies, clear priorities that help us achieve them, and new guidelines for the use of resources. Third, social and decision processes should ideally foster the creation and sharing of knowledge, provide an opportunity for conflicting values and opinions to be heard, bring about enhanced understanding, and develop lasting trust.

In short, these processes should be capable of clarifying, securing, and sustaining the common interest at relevant scales. How these processes will be structured in the future is still an open question, but the widespread support for the concept of sustainability opens up the opportunity to explore.

Conclusions

A complex interplay among leaders, bureaucracy, and context explains the patterns of conduct that we see in the leadership of the GYCC and other organizations and in the structure of natural resource management policy in greater Yellowstone. Leaders, of course, are individuals with personalities, beliefs, values, leadership styles, strategies for coping, and modes of discourse. Bureaucracies, designed to address certain kinds of problems using standard operating procedures, technical staffs, and existing programs, are good at routine tasks in stable contexts. But because the greater Yellowstone arena is full of dynamic, interdependent problems, its bureaucracies are having great difficulty coping. As for the context of greater Yellowstone, the natural resource management regime has been fairly stable, but many pressures are rapidly growing that may force changes. Overall, factors converge in the region's leaders, including those in the GYCC, who favor conservative, bureaucratic decision making; operate in a less than a fully contextual and integrative way; and give little attention to calls for innovation and systematic, active learning.

Greater Yellowstone can be considered a shareable resource, an inclusive common interest that must be managed with greater levels of cooperation, integration, and sustainability than ever before. The ongoing constitutive process in greater Yellowstone must mediate conflicting claims about competence over resources and decision making. As chronic problems continue to compound, demands will mount to adjust the constitutive process in favor of more inclusive interests.

Looking at greater Yellowstone functionally offers an understanding of its dynamics, especially against the backdrop of ongoing global dynamics. Jan Schneider, a legal scholar, notes that the world is a tight-knit set of ecological interdependencies, which condition and, in turn, are conditioned by human practices and institutions to satisfy bodily, psychological, and social needs and demands.[77] This calls for cooperation and integration.[78] Leaders must help control injurious use of shareable natural resources and facilitate their constructive, sustainable use.

Robert Keiter, an environmental law scholar, concluded that a "new era is dawning on the western public domain. We appear poised on the brink of the age of ecology in public land policy."[79] He said that how "this state of affairs has evolved, what it means, and where it may lead is a fascinating study in the power of ideas, science, law, and institutions."[80] A new direction in greater Yellowstone can emerge from the many accomplishments to date, the hard work of leaders, and the engagement and support of many people in the public that will put us on a genuine path toward sustainability. What will happen depends in large part on the region's leaders. The GYCC stands in a unique position to assume a strong leadership role and propel greater Yellowstone toward sustainability by elevating the debate, raising consciousness about the importance of the sustainability goal, and encouraging new patterns of democratic interaction that open up options for us all.

6 Improving Leadership

A leader's job is to be as contextual as possible in order to be effective in addressing the suite of ordinary, governance, and constitutive challenges that faces society. This is especially important in large-scale arenas like greater Yellowstone, which also has highly symbolic meanings for people around the world, because these three interactive levels of problems can have major consequences. But, says policy scholar Yehezkel Dror, leaders and society are too often unprepared to address these challenges, and the mechanisms of governance are in too many cases obsolete.[1] In these situations, dilemmas abound, opportunities to make gains go unappreciated, and political and bureaucratic cultures persist as obstacles to improvements. To meet these challenges, leaders must become successful networkers of people and multiprocessors of information. In the words of anthropologist Mary Douglas, leaders must function as an effective "thinking institution."[2] They are the "central minds" of society and should function together to address these interconnected challenges. Good leadership is a matter of skill in integrated problem solving, skills that can be increased, honed through practice, and used contextually.

In this chapter I explore options for leaders to develop skills to narrow the gap between current management policy and stated goals and offer a cluster of strategies to consider. These ideas are meant to be general enough to be applied in diverse contexts throughout greater Yellowstone or, for that matter, worldwide. Taken together, they can help leaders move beyond the constraints of bounded rationality (their own and others') as it exists today and overcome the limitations of less than fully effective organizations. Several strategies are recommended to encourage "out-of-the-box" thinking, practice-based action, and active learning. We know that such strategies work in actual practice.

Options for Improving Leadership

One of the most critical resources for addressing public problems is high-level leaders.[3] Ideally, leaders exist as a "community of good judgment," according to Ronald Westrum, an organizational sociologist.[4] Such a community should possess a culture that solves the right problems, it should be efficient at finding out what the real problems are, and it should be designed for rapid feedback and learning. Currently, many factors in greater Yellowstone—including the administrative style of leaders, bounded rationality, scientific management or positivism, the institutionalized mode of problem solving, goal displacement or inversion, weak decision-making processes, and reliance on bureaucratic standard operating procedures—all conspire to limit the good judgment that is needed. This may constitute the most serious obstacle to progress in greater Yellowstone because, as Dror noted, "the resulting inadequacies constitute one of the most difficult barriers to modernization, which frustrate many contemporary . . . policies."[5] Options for improvement do exist, though, including offering leaders opportunities to improve their skills, creating a "license to think," and getting clearer on goals.

Targeted Self-Improvement

Leaders can consciously and actively take themselves on as targets for self-improvement. Improving leadership skill is a broad notion that requires upgrading rational and extrarational capacities for management and policy making (by improving, for example, critical thinking and honing observational, management, and technical skills).[6] Improving their skills is vital because leaders set strategic direction and determine what and who will be supported through what kinds of problem solving and rewards. They also determine patterns of cooperation for managing resources that transcend jurisdictions, decide which practical demonstrations will be implemented, and shape the operational culture and promote or inhibit active learning.

The community of good judgment is composed ideally of the "central minds" of governance (that is, all the people in all the organizations engaged in cognitive processes about important choices and future-shaping capacities).[7] In greater Yellowstone these people work for diverse government and nongovernmental organizations. In a narrow sense, the target for improvement is people at top decision-making levels in government, people who play major roles in making critical choices. The GYCC is one key part of this smaller group, and it is the committee's stated goal, and thus its task, to bring about sustainable transboundary management policy. As such, its members need to be innovators involved in a creative process that will dismantle the old system or adjust it responsibly. Such a transition is a partisan process because the emerging system will be judged against the older one. Most current leaders have spent a lifetime building the existing policy system and developing skills and loyalties to maintain and enhance its arrangements; consequently, they must also terminate their old styles of leadership and loyalties. This is difficult, but there are ways for them to move forward.

Leaders can ask themselves what basic policy goals they are willing to recommend to responsible citizens. In approaching the policy problems of resource management, leaders can take the standpoint of identifying with the whole community and common interests, rather than with a single parochial interest, an agency bureaucracy, or a special group. New policies can help establish and promote conditions for human freedom and support basic institutions protected by the existing legal order.[8] Resource issues are important to all people (inclusive interest), although particular persons, groups, businesses, or states may have significant exclusive interests in the management of certain resources. In practice, this will require striking a balance among the competing claims of all people. To achieve this, leaders must be clear on their own standpoints, including their present skill levels, beliefs, values, biases, and other distinguishing characteristics.[9] They must account for these in their own decision making.

There are reasons why little is being done to improve leadership skills for resource managers and decision makers in general.[10] First is that current leaders are often not regarded as a legitimate target for active development, including basic and high-order skill development. Efforts to intervene with the "natural selection" processes (for example, recruitment, maturation, and promotion of leaders) that produce the political elite are seen as manipulative. Second is that the training of leaders constitutes a highly sensitive interference with the politics relative to the training of other professions (such as wildlife biologists, engineers, or technical assistants), even though the training of high-level leaders focuses in large part on highly controversial, nontechnical issues. Third is that active development of leaders is a difficult process under ideal circumstances, requiring great effort and a long-range standpoint. Few ready-made methods are available. Improvement requires active learning and smart trial-and-error insights. New types of teaching materials must be developed to help leaders advance their skills. Progress in the development of leaders is also difficult to evaluate and demonstrate. Moreover, concerted action to upgrade leaders' skills is in no way a panacea; not too much should be expected from it. Nevertheless, improvements, if feasible, can be a highly efficient way to help organizations achieve significant benefits.

Leaders can also work toward integrated solutions to problems.[11] They can do this, first, by using a logically comprehensive framework that invites, even actively encourages, integration of biological and social science, and also of general and local knowledge in their problem solving.[12] Second, they can integrate empirically derived case material about management problems into such a framework. This will help them check on what is known about the issue at hand and its context and learn the lessons of experience. Examining case material against recommended standards for decision making will also provide a reality check.[13] A third thing they can do is to seek integrated, win-win solutions, which will likely require new modes of cooperation, inclusiveness, and openness as well as redefining the context so that conflicting parties can satisfy their underlying value demands, if at all possible. This approach brings new practices into being, such as "prototyping" and other innovations.[14]

Creating a License to Think

Creativity is the only way out of the present dilemma for leaders and for everyone involved.[15] In short, leaders must create a "license to think," a means to encourage and facilitate policy-oriented creativity throughout the entire policy system.[16] This license must apply to themselves and everyone they work with. A few individuals in the greater Yellowstone region have already done so for themselves, but they do not currently occupy high-level positions in sufficient numbers to be felt in regional management policy.

Assuming that simple solutions exist to address the challenges, especially governance and constitutive ones, is a delusion. In reality, leaders are often hard-pressed to come up with practical management policies on some intractable issues (such as wildlife diseases in greater Yellowstone, large mammal migrations, or coexisting with large carnivores). The only alternative is to invent and develop new policy solutions and new capacities to implement them. Being innovation-ready is not something that just happens on its own. It depends on individual leaders from among members of the rank-and-file agency workers and from the public. It also depends on the advancement of new knowledge and theories, which sometimes run counter to established structures and processes, especially bureaucratic and disciplinary ones.

If present leaders assume that existing organizational and institutional cultures and structures are sufficient for creativity and that they are addressing all the challenges currently, then they are in serious trouble. Most current arrangements are much too constrained to permit the kind of creativity needed to secure a sustainable future for greater Yellowstone. The time has passed when recycling the same old partial methods and half solutions will suffice.[17] Our context or operating environment today is getting more complex, largely because of the forces of modernization driven by ever-accelerating science-based technology and by an expanding population. This fact is obvious even in the smallest, most remote town in the Yellowstone region. Modernization places new demands on leaders, including increasing complexity; accelerating rates of change; growing levels of aspiration; and changing political participation, communication, and quality of public policy making.[18] This is why leaders must be able to map the context realistically in preparation for making decisions.

Not only is creativity necessary, it is possible. There is great potential within the existing staffs of government, NGOs, and the citizenry to find better ways to address natural resource management problems. People across the board need to be freed to be creative, to take risks, and actively to learn and improve. Unfortunately, though, the region is littered with people who have tried to create solutions to problems without the support of leaders. Only high-level leaders can issue such license and support staff and others in their efforts to invent, develop prototypes, and in general improve problem-solving responses in a responsible fashion in the public interest.

Arthur Goldman, a philosopher of science, has described several components of problem solving that can be changed to open up creativity and discourse among leaders.[19] First, leaders can come to recognize the narrow way in which they define problems to themselves and then develop new ways that are not so limiting. Different conceptualizations may make it harder or easier to solve problems. The importance of "framing" problems should not be underestimated. As part of this effort, leaders can avoid committing themselves to initial representations and keep in mind that experts and novices represent problems differently. Second, leaders can become better at abstracting general ideas from particular circumstances. Some leaders are good at distilling "macrostructural" representations of problems and devising analogies or abstractions that can be crucial to solving problems. Breakthroughs often happen this way. Third, leaders can create the social settings in which creative problem solving can happen. Creativity and problem solving within teams generally outstrip those by individuals. Groups vary in their practices and structures; some actively promote creativity whereas others stifle it. Leaders can offer incentives for creativity and impose sanctions for behavior that is inimical to truth. Good leadership will embrace all these tasks.

A culture of creativity and active learning is needed in greater Yellowstone. This would give leaders options to accelerate positive trends that could move management policy toward transboundary goals and sustainability at a quicker, yet responsible, pace. Leaders should unite to encourage efforts to achieve a sustainable, free society that can live in a healthy environment everywhere. They can do this through their work of actively changing perspectives, strategies, and institutions toward this overriding goal.

Getting Clear on Goals

Clarifying goals, official and unofficial, personal and professional, is another promising means to improve leadership. Evidence suggests that both leaders and the constitutive process are partly "stuck or frozen" in greater Yellowstone, not knowing where to go or what to do. The struggle for goal clarity is evident. Without clear goals, it is impossible to find direction. In functional terms, setting goals can help create a social process whereby people can balance the inclusive competence of the larger community with the exclusiveness of local interests to secure the common interest for all. This balancing process must fairly produce and distribute both natural and human resources and the values produced by their management.[20] The challenge for leaders is to develop a functional understanding of goals, their role in the policy process, and how they might help balance diverse interests to serve the common good. This kind of orientation is difficult for many leaders, especially those in bureaucracies and ideologically driven organizations. It may require a significant reorientation for many of the region's administrative leaders.

This view of social process goes to the very heart of the constitutive policy process and governance mechanisms in our culture. Stated most broadly, the goal of our society's constitutive process is to increase the scope of individual choices now and in the future. Thus the constitutive process, in support of human freedom and democracy, should encourage the formation and efficient operation of communities, which can then arrange (and rearrange) and allocate (and reallocate) competencies relating to resources, people, and institutions as circumstances dictate. Needed change and improvement in greater Yellowstone can only come about if the constitutive process is rearranged accordingly. Successful constitutive process should encourage people and communities to secure the closest approximations to individual and community goals over time and through changing circumstances. The constitutive process should be set up and operated in such a way as to require the widest possible democratic sharing of power in decision making because it is these decision processes that produce and distribute all the values to people. A balancing of inclusive and exclusive competencies (that is, who has a say or who decides) is the ultimate goal of constitutive processes. It must be remembered that both inclusive and exclusive interests are in the common interest, if properly set (for example, proper oil and gas development by business produces the energy necessary for everyone). Leaders, who play a key role in this process, must remain flexible and help the constitutive process reappraise and reset the allocation of competencies over resources as the context changes.

The social process should permit and encourage democratic participation of all people who can affect or be affected by the making of decisions. It should reflect and be part of a system of public order that promotes human dignity and common interests and rejects special interests. It should facilitate establishment of, and access to, the constitutive arena by all citizens. It should enable more people to participate in and collaborate in decisions. The arena should be open and available to all participants with interests. If the arena is set up in this way to perform its societal functions, it will produce good constitutive process.

The social process in an open arena should be able to carry out all pertinent decision-making functions or activities. Information should be gathered, processed, and disseminated as needed. It should be available, open, and free to all. Debate should be balanced and inclusive. The rules (norms, laws, regulations) that have been decided on should stabilize people's expectations about what the future will look like, and they should be seen as fair. Implementation should be timely, open, dependable, and ultimately effective. It should be uniform and unbiased. Independent bodies such as universities, professional bodies, and the press, among others, should perform monitoring and appraisal, which should be flexible and should encourage constructive change. The social process should encourage persuasive over coercive strategies to address conflicts. It should promote strategies that are open, not secret, as well as fair, relevant, comprehensive, effective, and dependable. It should seek outcomes that are rationally designed to secure the community's common interests. It should seek to produce values appropriately and share them fairly.

Given these general, idealized goal considerations, leaders can generate specific propositions and actions tailored to fit real contexts. These can be derived in concert with the public in genuinely participatory ways. However, this is a difficult job. First, the context of goal setting is complex. Leaders are a part of community processes at international, national, regional, and local levels, and their activities have consequences, often conflicting, at all four levels. Leaders are not neutral or separate from these community processes; each of them is also a citizen of these various communities as well as a participant in their governing processes. Leaders have many value connections to these multiple communities. It is in this context that leaders must carry out their work. There are no value-free methods for leaders to use in setting and meeting goals. Leaders' actions, whether informed by knowledge or not, have value consequences. According to Myres McDougal and his colleagues, it is the obligation of leaders, supported by community resources, "not merely to relate law to some kind of policy, but rather, and further, to clarify and promote the policies best designed to serve the particular kind of public order they cherish."[21] There are always claims on leaders being made by various interests. A leader's job requires the kind of deliberate and thoughtful clarification of goals and processes necessary to learn about, address, and deal with these value dynamics in ways that serve common interests.

Goals must take into account the interdependence of people and their interests. Some people with a stake in greater Yellowstone spend their whole lives in the ecosystem and others live outside it; both groups are clearly involved in each other's affairs. These relationships are pervasive and growing deeper over time so that even small changes in one part of society—the economy, national security, the environment—can impinge on other people. The intensity of interaction and interdependence of people is growing dramatically. All the value dynamics that underlie social decision making—wealth, power, rectitude, respect, skill, knowledge, affection, and well-being—are affected by this growing intensity of interaction. These interactions find strong and clear, although occasionally dysfunctional, constitutive expression today. This is evident in most trends in greater Yellowstone—the industrialization of recreation, suburbanization, population growth (permanent and seasonal), depletion of resources (especially open space, wildlife habitat, solitude, quiet), and a deteriorating environment (such as air and water quality, biodiversity). These all make the whole social process more interdependent and more complex.

Balancing inclusive against exclusive interests can be difficult for leaders in such a setting. The terms of inclusive and exclusive interests are not fixed absolutely in reference to a particular resource or situation. The important job of leaders is to "achieve and maintain a continuous balancing of inclusive, partially inclusive, and exclusive interests which will serve the long-term aggregate common interests," notes Mahnoush Arsanjani, a legal scholar.[22] Leaders can begin by engaging themselves in basic goal clarification exercises as well as engaging the public constructively about goals.[23]

Upgrading Problem Solving

The first and foremost work of leaders is problem solving. How leaders think about and address ordinary, governance, and constitutive challenges determines whether management policy is a success or a failure. Options exist for leaders to improve problem solving, including being more formal or conscious in their approach to understanding, defining, and addressing problems.

Formalizing Knowledge and Skills

Generally speaking, people do not solve problems in a formalized or fully conscious way. The way in which people are conventionally trained to think and the standard operating procedures and existing programs under which they work largely confine problem solving to preconscious, rote styles. To the degree that this happens, more creative, better options are ignored. Leaders have the option to move beyond this conventional approach by culling the best from their already considerable experience and skills and formalizing it. They can do this by raising their consciousness and systematically making visible the present stock of knowledge and skills and how they are being used in problem solving. They can also become fully conscious of the constituent parts of the problem-solving operation and try to optimize these. Doing so can result in more successful problem solving.

There are different ways for leaders to conceive of the knowledge and skill set needed for effective leadership. One useful way calls for mastering the interrelated skills mentioned in earlier chapters—critical thinking, observation, management, and technical competence.[24] Critical thinking can be thought of as involving five problem-solving tasks that permit users to orient to problems realistically and efficiently, both procedurally and substantively. These tasks, introduced in chapter 2 and summarized below, are well known but not often talked about systematically or applied in fully conscious ways. They can help leaders to be procedurally rational. Observational skill involves finding problems or "targets" to focus on through the use of extensive (broad) and intensive (focused) methods. This can help leaders address the right problems. Management skill involves interaction with people, individually and collectively, through public education, diplomacy, speaking, and writing. This can help leaders facilitate problem recognition and solution. Technical skill involves knowing the distinctive skills of a profession (such as geographic information systems, participatory assessments, or ecological methods). This can help leaders stay abreast of developments. Applied systematically, these four interrelated skills can lead to genuinely interdisciplinary problem solving. Its promise in the GYE is that it can provide a logical, comprehensive approach to meeting the many diverse problems confounding sustainability of the environment. Using multiple methods to understand and solve problems is essential, too. Diverse biophysical, social, and local knowledge methods all have something to contribute. Taking stock of these skills in any individual leader or leadership team is an important first step in improving leader performance.

To compound matters, problem solving is often done preconsciously and may be retarded by a host of psychological, sociopolitical, conventional, and competitive factors and forces. These barriers prevent people from using their knowledge and skills in effective ways.[25] Both leaders and rank-and-file workers must try to overcome the barriers posed by ideologies, objective-subjective tensions, personality dynamics, and different styles in perception, memory, reasoning, and personalities, all of which may pull against each other to limit rational problem solving. Group decision making can also be affected by these same problems. People are thrown together, often temporarily, to solve problems. What results is a mix of people using different forms of reasoning, directed by different ideologies and value outlooks, to capture problem definitions and promote their preferred solutions. The result often does not satisfy anyone nor does it lead to effective solutions.

It is clear that conventional problem solving itself can be a major problem. One of the first things leaders in greater Yellowstone can do is to explore the knowledge and skills presently available in their own operations, their staffs, and their working relationships and, if they find shortfalls, to upgrade as needed. Developing a conscious, formalized understanding of problem solving opens up many promising options to be more effective at the task.

Becoming Fully Problem Oriented

To improve their effectiveness, leaders can employ a practical frame of reference for their problem-solving operations. Rational problem solving, that is, striving to be both procedurally and substantively rational, continues to recommend itself over irrationality. A person is "purposively rational if he considers the goals, the means, and the side effects, and weighs rationally means against goals, goals against side effects and also various possible goals against each other."[26]

The preferred approach recommended here provides a practical frame of reference that puts the social and decision process within the scope of potential understanding and guidance. It also provides a useful analytic framework to guide problem recognition and solution in realistic ways. It helps users perceive all the important features of the social process in relation to the entirety of events in the situation. Finally, it invites and employs diverse methods. Problem solving requires much more than the derivational logical exercises of scientists, lawyers, or administrators; it requires the whole complex of interrelated activities described in detail in chapter 2 and appendix 5. Gaining knowledge and skill in this approach is both possible and highly practical.[27]

Attending to Higher-Order and Basic Tasks

Leaders have the responsibility to gear up for higher-order tasks and engage themselves in basic philosophical questions that bear on the problems and solutions under consideration.[28] Such efforts differ from ordinary tasks, which encompass the day-to-day work of leaders and the routine provision of services by bureaucracies, which serve only to maintain existing arrangements and satisfy small-scale needs. More efficiency in ordinary work does not constitute a high-order or basic "improvement," although such efficiencies are the focus of many schemes purported to improve governance and constitutive problems. The demands of a challenging context call for more attention to the higher-order tasks, those that Dror says "involve significant efforts to shape the future."[29] In the context of greater Yellowstone, these tasks call for leaders (and society as a whole) to look at the heart of our constitutive process—the concept of society's relationship to nature. It seems obvious that improving that relationship will put the region on a solid trajectory toward sustainability. For leaders to be able to address these vital higher-order tasks, certain prerequisites must be met. Among these is a strong leadership interest in the kind of goal clarity that can create a common vision for the future, drawing us all toward sustainability. Perhaps hardest will be solidifying the long-term political will in the region to take on these tasks.

More specifically, these higher-order tasks can be addressed by using scenarios and policy exercises.[30] Garry Brewer, who is skilled in these methods, notes that such tools allow users to think creatively about overcoming complex social and environmental problems that interact and evolve over large time and space scales. Modeling can be used, too, to help synthesize vast quantities of scientific and social knowledge. Many techniques exist for this purpose.

But all of these methods require as a condition of success that leaders be conceptually and practically clear about what "change" entails. Noel Tichy, an organizational change specialist, says that there are three major foci of change—power, values, and technical aspects of work.[31] These refer to the political, cultural, and technical dimensions of change, which can be targeted through organizations, social networks, and management, respectively. Levers to implement these changes include prototyping, leader development, and active learning.[32]

In sum, many options are open to leaders to formalize their knowledge and skills about problem solving, to become more explicitly problem oriented, contextual, and multimethod, and give more attention to higher order tasks of leadership. Actively exercising these options can help improve regionwide policy for managing resources.

Creating Capacity

Successful leaders depend in large part on the capacity of available individuals, groups, and organizations to help identify and solve problems. Capacity, the "ability of actors (individuals, groups, organizations, institutions, countries) to perform specified functions (or pursue specified objectives) effectively, efficiently, and sustainably," always seems to lag behind challenges.[33] If capacity is weak or absent, then it can be developed through efforts made by the actors themselves, especially leaders, to achieve or strengthen their ability to perform the functions in question. "Capacity can be built through a wide range of activities undertaken by the various actors involved, and is not a restrictive or prescriptive term."[34] Among such activities are education and training through workshops, seminars, lectures, and short and long courses. These activities can result in adoption of new knowledge and skills for individuals and improved observation and management practices.

Staffing and Partnerships

Leaders throughout greater Yellowstone have diverse staffs and a network of potential partners who can help address problems. Throughout the region there is a combined staff across all organizations of a few thousand individuals who focus on finding and solving problems in natural resource management. Some leaders can draw on hundreds of staff members and millions of dollars, but most are more limited in capacity. There are also many potential partnerships in place or being developed to make up for shortfalls in staffing, knowledge, or skill within any one organization. However it is done, getting more help is always a good option. Existing capacity should be compared to the challenges at hand in order to identify shortfalls and then to build the needed capacity.

There are two main approaches to building capacity.[35] One approach, based on philosophies that seek generalizable laws and broad problem solutions, is to build capacity for "hard" systems, that is, scientific and technical expertise (in positivistic specialized knowledge and skills). The other approach, emphasizing solutions that are context specific and outcomes that may need to be renegotiated over time, is to build capacity for "soft" systems, those involving diverse players and different knowledge systems in policy making, including pre- and postpositivistic knowledge systems and local knowledge and skills. Currently, the emphasis in greater Yellowstone is on the capacity of hard systems. In our system of public order, government has a capacity-building role to play for itself and for the services it provides to citizens. This role stems from a long history of capacity-building efforts at both the state and federal levels in cooperation with citizens, associations, and other organizations. Government and its leaders, however, can be more active enablers of the community than they are at present.

A key role that can increase capacity in a system is the "advisor," someone who is good at effective problem solving, synthesizing methods, and other professional support services.[36] Ideally, leaders need to surround themselves with an informal circle of advisors who are skilled in these services. Sometimes they must rely almost exclusively on these advisors, who can help them cope with the rush of affairs, keep an eye on the larger social structure of the policy system, and consider the viewpoints of diverse individuals in the context. These people can have an enormous integrative influence on leaders. Such advisors should be willing and capable of taking on part of the burden of leadership: there is nothing like being on the front line to make an advisor understand the hard realities. A really useful advisor needs to look beyond immediate views and concerns to consider the longer-range and bigger-scale prospects. Advisors may not be needed if leaders have outstanding judgment and clear ideas in every field of policy, but this is seldom the case, and most of the time leaders need the help of advisors.

Partnerships are another important capacity-enhancing tool. Partnerships within and outside of government can go a long way in meeting shortfalls in staffing capacity. They can build complementary sets of knowledge and skills for particular or broad problems, especially when resources are limited. Presently, the value of partnerships can be seen in the proliferation of committees and subcommittees and the many growing networks in the region. The region's leaders can help bring a more widespread and active use of partnerships.

Functioning in High-Performance Teams

High-performance teams demonstrate results that are consistently superior over time.[37] Leaders can encourage their staffs and partners (including committees and subcommittees), wherever feasible, to function as high-performance teams. But such teamwork rarely just happens. David Hanna, an expert on the functioning of teams, said that people "seem to lack the essential ability to work together effectively to solve critical problems" without the application of special effort.[38] His research showed that high-performance teams exhibit an ability to read the operating environment (context) well, including positive feedback. Teams are also able to sense new opportunities based on existing skills. Their flexibility allows them to rearrange operations, balance tasks, change individuals, group together core processes, and effectively support new purposes. They show a sense of ownership among leaders and workers alike; all members seem to have a stake in the success or failure of the effort, and each sees how his or her efforts make a difference. High-performance teams are also successful at self-regulation.

Principles and practices of high-performance teams are known and are based on an understanding of human behavior. Among the characteristics of effective teams are development of specific performance goals and objectives and close coordination of activities among members to attain their common goals. These goals must be elevating and empowering. The team should be "results" driven and have a structure to achieve results. It should have competent members who are unified and committed and who share a collaborative climate and high standards of excellence. It needs principled leadership as well as external support and recognition. Effective teams can improve members' skills, increase learning, enhance thinking, and boost productivity. High-performance teams, however, cannot produce quick fixes. They require time and work to achieve results. Teams work for a common purpose and struggle against the flow of conventional activities, especially complex, bureaucratic ones.

Creating and operating high-performance teams usually requires knowledge of the workings of open organizational systems.[39] The dynamic processes of a high-performance team or any other organizational system include inputs, transformation, and outputs, which must be matched and managed for the team or organization to work well. There are great opportunities to advance the sustainability goal in the Yellowstone region by creating and supporting high-performance teams to complement the current collections of technical professionals and administrators.

Using Problem-Solving Seminars

Within bureaucratic agencies and other relatively large organizations (such as the Greater Yellowstone Coalition) and in small groups like the GYCC and even local NGOs and businesses, leaders in greater Yellowstone face the challenge of finding and using effective problem-solving routines. The problem-solving seminar is one tool that has yielded proven benefits.[40] This group-based, decision seminar strategy, originated by Harold Lasswell, has proven to be the perfect mechanism for "blend[ing] wisdom and science, for balancing free association and intellectual discipline, for expanding and refining information, and for building problem-solving cultures that balance 'permanent' with 'transient' membership, thereby remaining open to new participants and to fresh ideas while developing a capacity for cumulative learning that refines, clarifies, and simplifies."[41]

This method is effective because it focuses on several key requirements essential in any successful problem-solving situation. It demands a contextual approach to problem solving, permitting participants to go back and forth between the parts and the whole and traverse the past, present, and future. It requires multiple methods that allow participants to use a variety of approaches to problems and invite in all the disciplines. It helps participants to be clear in stating purposes, goals, or objectives. The decision seminar permits participants to use all the problem-solving activities in a way that encourages new understanding and at the same time promotes consensus. It requires a problem-solving strategy embedded in a genuinely problem-solving group culture and a well-grounded conception of social process, values, and decision process. The problem-solving strategy must be carried out in a problem-oriented analytic sequence, which acts as a guide or an investigative checklist to problem solving. Issues of content (such as natural resources) and procedure (human interactions) must be attended to simultaneously. A focus on integration is essential. A decision seminar also requires resources in order to work, among them a permanent core group, a permanent location, outside experts, record keeping, research, and appropriate support services (such as audiovisual aids).

In short, the decision seminar method fosters configurative or integrative thinking.[42] It is a rational method of deliberating and decision making with significance for anyone who works with policy problems. According to Leonard Cunningham, an educator, its skilled use can "broaden the base of input for decision making, improve intra-system communication, provide a uniform, structured, and disciplined approach to problem solving, decrease reliance on reactive decision making, develop a permanent leadership core of individuals at all levels within the system, and reduce the negative effects of hierarchy and bureaucracy."[43]

Conclusion

Other options to improve leaders also exist in addition to staffing and partnerships, high-performance teams, and problem-solving seminars.[44] These include better understanding of the relation between science and policy, public education and policy, and a host of other leadership improvement techniques that have proven useful in other settings.[45]

In spite of decades of hard work and many notable successes achieved by greater Yellowstone's leaders in and out of government, the problems—ordinary, governance, and constitutive—are growing faster than leaders' ability to deal with them and the policy system's capacity to respond. The region's leaders today have an opportunity, particularly the GYCC in its unique role as a high-level, interagency committee, to close this gap as effectively, efficiently, and equitably as possible. An enormous amount of experience can be brought to bear to help the region's already knowledgeable leaders, and many proven concepts as well as exciting new ones are available within and beyond Yellowstone that can also help. The model of a healthy greater Yellowstone ecosystem guided by enlightened management policy is a widely popular idea. This is the time for leaders to build the necessary coalitions, overcome dissension, and integrate fragmentation into a vision and action that moves us all toward sustainability.

7 Improving Management Policy

Policies that are developed for managing natural resources everywhere are based on what people value and the capacity of their organizations to meet those value demands. In other words, policy decisions are part of a human social process, which, in turn, relates to the decision-making process and institutional behavior. As a result, among the foci for making improvements should be the social and decision processes. The same social process that creates problems must be used to solve them. According to political scientist Charles Lindblom, "A fundamental, lasting, long-term requirement for good problem solving is . . . to create opportunities for collective action that did not exist before."[1] The way to achieve improvements in policy is to upgrade decision making and then test outcomes to see if they serve common interests. This strategy is an interactive, iterative one of action and feedback. Another way to make improvements is to organize people so they can interact more constructively, which requires that we attend to how arenas are set up and that we become more inclusive and contextual in our problem-solving efforts. Another improvement is to apply the concepts of active learning, including prototyping and proactive learning. Recruiting and training new leaders are also essential to make gains, as are adaptive leading and managing. Ultimately, developing sustainable policy on an ecosystem scale requires recognition that solving public policy problems cannot be reduced simply to scientific management, technical issues, or to some set of ordinary problems.[2]

In this chapter I offer options to improve management policy in greater Yellowstone by focusing on decision making, avoiding common pitfalls in decision making, organizing the arena better, adopting active learning, and managing adaptively, all in the interest of identifying and supporting common interest outcomes.

Focusing on Decision Processes

When leaders understand that improving policy means upgrading how decision process functions, then widely recognized standards for good decision making can be used to appraise existing decision process. This can be done regardless of the level of decision making or the subject of the decisions.

Pitfalls in Decision Making

As we saw in chapter 4, the decision process is the way societies resolve political differences or at least keep any conflict that arises from differences within acceptable bounds. Improving decision process is one of the keys to better policy making for managing resources. There are always claims and counterclaims being made by special interests about the adequacy of decision processes, regardless of their content or the situation. It is the job of leaders to address the inherent value dynamics of these competing claims. Leaders also have a key role to play in helping communities avoid the many common difficulties that tend to plague every function or phase of decision process.[3]

One pitfall that often develops in the initial phase of the decision-making process is "delayed sensitivity," in which perception of a problem arises only after the problem has progressed to the point that harmful effects are widely felt. Leaders can avoid this problem by paying close attention to changing contexts, making projections about developing issues, and anticipating problems before they get out of control. Another potential danger is biased initial problem definitions, wherein one special interest group puts forward a problem definition that favors its own interests and its view of the problem, thus creating a definition that fails to capture the full and true nature of the emerging problem. If leaders can account for competing problem definitions and sort out how different definitions serve different interests and values, then they can find ways to achieve common interest outcomes.

In the estimation phase of decision making, inadequate analysis of the problem may create difficulties. Vital data may be lacking or analysis of trends, conditions, and projections may have been only partially carried out. Another problem that may emerge is endless calls for additional study or insisting that further research on the problem is needed, as a way of buying time. This is a common tactic of people who oppose the emerging policy picture and problem definition. Since they do not accept the problem definition or want to take action on it, their policy is to maintain the status quo.

Among the risks in the selection phase is poor coordination in government decision making, as, for example, when several groups, who may not be aware of one another or do not communicate well, work on problems simultaneously. As a result, they fail to build a common understanding of the problem and its potential solutions. Overcontrol is another problem, wherein groups may respond to problems by imposing greater controls on everyone involved in an effort to force their definition of the problem to the forefront. This leads to bureaucratization and sometimes paralysis or gridlock as the conflict grows and people spend their time blocking each other's efforts.

The implementation phase of decision making may be afflicted with "benefit leakage," whereby most of the expected benefits are lost because of bureaucratic inefficiencies and other reasons before they reach their intended beneficiaries. Another implementation problem occurs because of the inherent limitations of government—size, inertia, political interests, conservatism, and bureaucratic features. Leaders also need to steer clear of poorly coordinated implementation. Often bureaucratic overcontrol, rivalry, and exclusion of key parties can lead to muddled implementation of policies and programs.

In the evaluation phase of decision making, insensitivity to criticism is one trap that can short-circuit progress. Critics may call for improving a policy, but government often simply ignores their input, regardless of its merits. Another issue is failing to learn from experience. Organizations, whether government, nonprofit, or business, sometimes repeatedly respond to new challenges with the same old programs, approaches, and techniques.

The termination phase of decision making is also subject to unique shortcomings, among them pressure to continue unsuccessful policies. Even poorly performing policies or programs that have outlived their usefulness may benefit someone, who then clamors for the policy or program to continue. Failure to prepare for termination is another peril; groups may fail to appreciate and prepare, early in the overall policy process, for the difficulties of terminating a policy that has outlived its usefulness or was flawed from the outset.

These hazards come about for many reasons, including structural, informational, and shared motivational factors.[4] The structure of decision processes is determined in large part by bureaucracy, which is ready-made to succumb to these pitfalls because of its relatively closed nature.[5] For example, some agencies are interested in maximizing internal control of their employees and maximizing advantage relative to other participants in their operating environment. Thus agency behavior closes down information flows, estimation of problems, time to consider, implementation opportunities, monitoring and honest evaluation, and effective termination.

The Forest Service, Fish and Wildlife Service, and National Park Service all operate as bureaucracies that control and limit internal creativity and the way decision processes are carried out. Information is often heard by leaders in highly selective ways and is filtered through shared frames of reference, which in turn influence decision processes. Motivational factors in these agencies include "hidden" value preferences that are implicit in all organizations. For example, one motivation that might override other decisional goals is sustaining and enhancing the bureaucracy's welfare or the jobs of particular bureaucrats.[6] Such motivations serve as a primary reference for group members and may have a powerful influence on leaders and workers alike.

Many of these factors increase impersonalization, internalize agency values, and displace formally expressed goals.[7] This, then, often limits the search for options or innovative methods of decision making and introduces a corresponding rigidity of behavior and a greater difficulty in dealing with the public. As Herbert Kaufman said about the Forest Service in 1959, an agency may select and train employees who have the "will and capacity to conform" to what the agency wants done.[8] Kaufman detailed the Forest Service's socialization process that evolved to produce compliance and loyalty. For instance, frequent transfers keep employees' loyalties to the agency rather than to the locale where they are working. A similar process exists in other agencies as well. Terrance Tipple and Douglas Wellman replicated Kaufman's study thirty years later and concluded that although there is rhetorical emphasis on responsiveness and representativeness, the Forest Service retains its old emphasis on scientific management efficiency and patterns of loyalty.[9] Although the National Environmental Policy Act (NEPA), the Endangered Species Act, and other pieces of legislation have forced the agency to engage in broader interaction with the public and the external environment, the later researchers concluded, "While much of what Kaufman found is still present, it exists in a more complex and diverse setting."[10] This pattern is evident in all the agencies in greater Yellowstone in varying degrees. Leaders who know how to avoid these potential difficulties by testing the quality and outcomes of the decision process will be essential to finding sustainability in greater Yellowstone.

Judging Common Interest Outcomes

Public policy for managing resources should serve common interests, which policy scientist Ronald Brunner says are at stake "whenever people who act on their perceived interests also interact enough to form a community around an issue."[11] Although never assured, the common interest can be achieved through high-quality decision processes. There are tests—at least partial tests—that can be used by leaders and others to help make qualitative distinctions about whether any particular decision process tends toward the common interest. These three partial tests of the common interest—*procedural, substantive,* and *practical*—are based on ideas about jurisprudence for a free society first articulated by Harold Lasswell and colleagues decades ago.[12]

Among the considerations in determining whether a policy serves common interests are that some interests may be inappropriate or invalid, that sometimes individuals misperceive what is in their own best interests, and also that individuals might reconsider their special interests through social process. The many individuals in the larger community share interests, which are not fixed and should never be assumed but can be made clear through the decision process. They can be secured and sustained with the help of high-quality leaders and an enlightened citizenry. There is no infallible or objective way or formula for assessing the common interest in any particular case—smart, experienced people of good will can and do arrive at different conclusions about the common interest—but these three partial tests can be used by all people to help determine the quality of decision process outcomes.[13]

The procedural test asserts that it is in the common interest for the decision process to have inclusive and responsible participation. This is consistent with our democratic form of government. Brunner says that, when applying this test, leaders (or analysts) should first observe closely to make sure that the effective participants, both official and nonofficial, actually represent the larger community as a whole. If that is not the case, then the decision process is not likely to reflect the values or interests that are not represented. Second, leaders should ask whether the effective participants are responsible, that is, are they willing and able to serve the whole community and can they be held accountable for their decisions. If the answer to either question is no, then they may not be serving the whole community. Instead, they may be special interest participants, which may be contrary to the common interest.

The substantive test rests on recognition that the central focus in common interest deliberations should be community members' valid (supported by the evidence) and appropriate (timely) interests, as measured by the values they espouse. This too is consistent with our form of government. In applying this test, according to Brunner, leaders should make sure that a person's or group's expectations are backed by solid, realistic evidence that is available to all people. Sometimes people seek values (i.e., self-interests) that are invalid because there is no supporting evidence to justify their demands. In such cases, the "evidence" offered is fabricated, highly distorted, or incomplete and used to support narrow special interest. If the evidence is lacking, then that interest may be dismissed as invalid. Sometimes, perhaps often, people assume that their interests are valid without bothering to see if the evidence backs them up. Leaders should also consider whether a particular value demand is consistent with larger, more comprehensive goals of the community. If it is not, it is a special interest rather than a common interest. Another way to look at the substance of the claim is to see whether other people who are known to be well informed and to represent the whole community have accepted and approved of the demand. If they have, then it can be considered to be in the common interest; if not, it is likely a special interest.

The practical test also rests on people's experience. To apply this test, leaders should realize that even inclusive, valid consideration of common interest might produce a policy that is inconsistent with people's experience. In short, the whole community, just like an individual, may misperceive what is in its interest. If the policy, as implemented, turns out not to be in the common interest, leaders should consider what can be done to open participation up so that it is more representative of the whole community and addresses the full range of interests, both the obvious and the more subtle ones. They should also ask, in justifying an existing policy, which interests are no longer valid and whether any emerging interests were left out of the original policy and should be integrated into new efforts. This test focuses on improving matters, not perfecting them.

Even if these three tests are largely met, it does not, of course, guarantee a common interest outcome that will be recognized and supported by all or most people. But being aware of these tests and striving to pass them can significantly help address complex policy dynamics for the better.

Organizing the Arena

Problem solving, cooperation, and coordination can happen if people can be brought together in productive ways. This means that the zone of interaction must be organized effectively. Well-organized arenas empower people, foster civic-mindedness, and promote joint projects and cooperative deliberation. In greater Yellowstone there are many examples of people organizing themselves—the Greater Yellowstone Coalition, the Yellowstone Business Council, science conferences, the Sustaining Jackson Hole project, the Trapper's Point citizen group concerned with wildlife migrations, and the Jackson Hole Conservation Alliance, to mention only a few. People are trying to resolve Yellowstone's challenges by creating effective arenas for themselves, often independent of government. Without an adequate arena, it is impossible to debate the nature of the challenges, decide what to do about them, and implement and monitor solutions. Although there are some effective organizations in greater Yellowstone, the present arena in the region is problematic. To address the arena's issues, leaders can help to link people, resources, authority, and control.

The greater Yellowstone arena consists of both the geographic area and the human landscape, and it deals with diverse perspectives, varying time scales (e.g., from the incremental nature of species decline to the rapid nature of property transfers or institutional reorganization), and competing interests. At present the arena lacks efficiency, access, and a resolution function for its many problems. One key to understanding the context of human interaction in Yellowstone is to understand the constructed quality of the arena itself. Greater Yellowstone exists as a construct of human imagination to meet the perceived needs of people and groups. This construct is a product of the different histories of people, the geographic region, and the participating institutions and social groups. The term *greater Yellowstone* is not recognized officially in United States law nor does it appear on official maps. The point here is that the term is simply a conceptual and management category created by people to refer to the aggregate of lands where various organizations are working around the somewhat serendipitous founding of our first national park. The trend toward managing greater Yellowstone as a single unit is gaining momentum because an arena is organizing itself around the concept of an interdependent ecology and community. If this trend continues, a decision process for the allocation of resources and values will become clearer with time. The constructed nature of the arena affects the kinds and quality of decision process that occurs there, even though few people recognize, study, or inquire about the properties or adequacy of this evolving arena and even fewer seek to manage the arena for optimal democratic participation and sustainability.

The kind of arena we create can help or hinder us in achieving common interest goals. As the greater Yellowstone arena organizes itself for debate, decisions, and management and the decision process begins to make it more visible, many institutions and social groups are gravitating to this arena for various reasons. New NGOs are appearing each year, for example. In order to understand the Yellowstone arena, we can try to understand the relations among these organizations and the ways they interact in the decision process. Leaders can promote this understanding by using a richer system of functional and out-come analysis that will enable them to cultivate an arena best suited to (and adaptable to) the context and the challenges. One single arena will probably not be adequate; a structurally flexible arena can provide a wide variety of smaller constitutive arenas that can be made to fit the diverse requirements of the many participants in Yellowstone manage-ment policy. The arena should be congenial to understanding and managing constitutive decision making in the common interest. Lead-ers can set up and maintain the arena through organization, meetings, deliberations, and actions. They should consider varying the present composition and role of people active in the arena and enabling new participants to collaborate in at least some decisions, including a more diverse array of professionals and the public. Leaders need an effective arena within which to work in order to minimize miscommunication with the multiple participants and the kinds of problems that typically flow from only formal bureaucratic relations.

Ideally, a greater Yellowstone arena can be structured, that is, es-tablished and readjusted, to meet the changing realities of the context, especially the valid claims of participants. This ought to be an ongoing process that is responsive to the interaction of constitutive and com-mon interest goals (e.g., Park Organic Act, Endangered Species Act, NEPA). It must also be responsive to changes at national and global scales, including global climate change, international biodiversity con-vention, and similar events and processes. The arena must be open to access by all those who want to participate. Effort should be made to optimize interactions so that the goals of the body politic can be achieved. Flexible strategies are required to balance centralized against decentralized aspects of the arena, organized against unorga-nized, specialized against nonspecialized, and temporally continuous against discontinuous.[14] This requires leadership and good manage-ment.

Functionally speaking, arenas should provide for the performance of all of the authority functions needed to promote the resolution of controversies.[15] This raises many questions about the adequacy of the present arena in greater Yellowstone. For example, who is authorized to participate? Who has admission to the arena? How are participants

regulated? Do rules or prescriptions exist about how the arena operates economically in terms of balancing or maintaining the security of
the larger community and promoting genuine self-direction in lesser
communities in the arena? Do these rules allow for change and adaptation? Are the rules and procedures about membership in the arena,
representation, and credentials stipulated and clear? Are they compat-
ible with easy access by all interested participants, or do they exclude
or subordinate some participants? Do they create controversy and
continuous tension? Is provision made for the reciprocal recognition
and protection by government of the private associations that seek
specific values, such as wealth or enlightenment? Do individuals have
open access to governmental arenas? These are some of the many
questions that leaders and others can be thinking about as they organize arenas.

The challenges confronting greater Yellowstone can be better addressed if participants are more civic-minded and if they are empowered to help solve problems. Effective participation is not only very
helpful in creating consensus, it can be highly rewarding to individuals.[16] For this to happen, however, people must be included in the decision process. Two types of analysis can facilitate any inclusive effort
and give leaders a good picture of the actual dynamics at play in the issues before them: *stakeholder analysis,* which seeks to clarify relationships and types and levels of power that exist among various actors,[17]
and *threat analysis,* which seeks to identify those activities of human
or natural origin that cause significant damage to resources or are in
serious conflict with the goals and objectives for managing and administering the area.[18]

There are many ways to think about participation and inclusivity.
Michel Pimbert and Jules Pretty, international development specialists, offered a typology of seven levels of participation ranging from
minimal to maximal participation, which was introduced in chapter
4.[19] Most of the agency-directed public participation in greater Yellowstone at present is at the lower end of the scale, although some citizens, of course, are self-mobilizing, perhaps in response to the limited
opportunities for meaningful participation through government-level
forums. If leaders in greater Yellowstone want to be more inclusive
they can move to higher forms of participation.

In the lowest level, participation and cooperation can be "passive,"
wherein people participate by being told what is going to happen or
what has already happened. People's responses are not taken into account. Second, participation and cooperation can take the form of
"participation in information giving," in which participation consists

of responding to questions from researchers and project managers via questionnaires, surveys, or similar approaches. People have little opportunity to influence proceedings, findings, or project designs. The third level is "participation by consultation" wherein people may be consulted and external agents may listen to views, but the external agents define both the problems and the solutions. This type does not concede any share in decision making to the public, and the external professionals are under no obligation to consider people's views. Fourth is "participation for material incentives," wherein people participate by providing resources, such as labor in return for food, cash, or other incentives. Fifth is "functional participation" wherein people participate by working to meet goals predetermined by decision makers. This kind of participation takes place near the end of decision processes after someone else has made all the important decisions. Sixth, participation and cooperation may take the form of "interactive participation," which involves people actually participating in joint analysis. This can lead to action plans and formation of new groups or strengthening the capacity of existing ones. In this type, groups take control over local decisions and therefore have a stake in maintaining structures or practices. Finally, participation and cooperation may be "self-mobilized," wherein people participate by taking initiatives on their own, independent of external agents or institutions, to change practices and institutions. Self-initiated mobilization and collective action may or may not challenge existing inequitable distributions of wealth and power.

Participation requires a forum or arena through which civic deliberation can take place and citizens and agencies can interact productively. These require the higher types of participation that Pimbert and Pretty talked about. For any forum to be useful it must promote communication, mutual understanding, and active participation in problem solving, all the while remaining legitimate and keeping a sociopolitical structure that is open and publicly supported.[20] The potential and actual benefits of these more participatory arenas are well known.[21] They solve problems and they recharge the public's trust in government, affirm that government is responsive to the public's needs, and foster active and useful democratic citizen input. They "encourage people to think of nature in new ways."[22]

One study that focused on citizen involvement in the Yellowstone region was graduate student Melissa Frost's analysis of the arena of wildlife disease management and its ongoing cycle of endless conflicts.[23] She recommended deliberative civic engagement as a way to disarm public distrust and discontent with government and actively

involve citizens in decision making. Such new efforts operate to enhance dialogue and cooperative enquiry into the nature of problems,
and they tend to enhance social capital by developing trust, reciprocity, and resiliency.[24] Many other people have also offered recommendations for civic deliberations.[25] Truly integrated solutions are hard to
achieve in practice but are worth striving for.

Active Learning

Policy improvements can also come about when participants become
more active learners. Most of us learn and change our behavior based
on our experience, but we learn at the individual level and few of us
actively carry out systematic reviews of our experience. Few people set
out to learn as much as possible from their experience. Few set active
learning as a personal goal.[26] We can all become more actively involved, looking at our operating assumptions and working to develop
insightful reflection as a basis for learning and improvement. Learning
occurs at multiple, interrelated levels—individual, organizational, and
policy system.[27] It takes place most often at the individual level, and
lessons at that level can be, but typically are not, successfully trans-
lated into organizational or policy system learning.

There are many tools available to help leaders and others learn actively, among them prototyping and case-based learning, or harvesting
the lessons of experience. This is a way to examine past experience,
find lessons, and diffuse and adapt them to future work. These tools
can be helpful in promoting practical problem solving and proactive
learning.

Prototyping

Prototyping is an innovative, yet proven, strategy to enhance performance in the kind of complex context that exists in greater Yellowstone today. A prototype is a model, official or unofficial, from which
something can be learned or copied. It serves as an exemplar or arche-
type. "Hence an aim of any prototypic study is to devise a better strategic programme," according to Harold Lasswell, one of the early
proponents of the method.[28] Because prototypes are not fixed in advance of beginning a project, they encourage ongoing learning and
creativity and can be copied and incorporated into other ongoing
work. Prototypes lie halfway between tightly controlled experiments
done by scientists and interventions by high-level administrators or
others (in the absence of experimental controls). They also serve as

case studies in the sense that they take note of actual, real-life factors that cannot be controlled.

A prototype is not the same as a pilot study, which is often planned in detail in advance. Although the objectives of a prototype are usually explicit, many ambiguities may remain. Prototyping should be kept in the hands of people who are solidly committed to knowledge and skill improvements. Most useful in situations of complexity and uncertainty, it is a positive step to improve professional and organizational performance. In order to work, first, there must be tentative consensus among participants to try it. Second, the context must be ripe, although resistance is to be expected. Third, because no blueprint exists, the process must be kept open and creative. Fourth, first-class professionals are needed, people who are respected for their ability to advance knowledge and practice. Fifth, a strong incentive is needed. Sixth, "politics" or destructive competition over values must be avoided. Finally, there may be people who oppose it, usually for ideological reasons, on the basis that it has "never been tried before," so participants must be committed to making it work. The transferability of prototypes to other contexts or arenas is always an issue. In rare instances, whole programs can be transferred into similar contexts, although usually only selected elements will be transferable.

This approach has been used in several instances in greater Yellowstone. Among the relatively successful prototypes are the Beaverhead County Partnership, Madison Range Landscape Assessment and Adaptive Management Project, the Henry's Fork Watershed Council, and the Greater Yellowstone Coalition Stewardship Program.[29] For example, Steve Primm and Seth Wilson, both grizzly bear researchers who work closely with ranchers, reviewed experience and cases for lessons in grizzly bear management.[30] Among their lessons is that people who have lived with bears have valuable insight and practical knowledge that should be used by managers, that these cases offer good prospects for designing innovative programs, and that capitalizing on such cases requires a systematic approach to understanding social context and involving local people in planning and research. Prototyping has been used in many other natural resource settings.[31]

Harvesting Experience

Finding lessons in the experience of actual cases is another key way to improve policy for managing resources. Treating each management problem as a case study shifts focus away from its technical, scientific, expert-driven aspects and integrates the technical, political, and cultural dimensions actually involved in each case.[32]. To understand cases realistically means identifying all the people involved, directly and indirectly, and the nature of the social and decision processes that make it up. Without a problem-oriented, contextual understanding, it is easy to ignore, overlook, or misconstrue important details, and as a result lessons go unlearned or the wrong lessons are learned and mistakes are repeated. In too many cases, legitimate issues and participants are left out of the "solution," or clear and easy options are passed over.

There are established case study methods for the "systematic recording of an event or series of events with the objective of learning from that event."[33] The case method is used in training doctors, lawyers, and other professionals, and according to Jeff Romm, cases offer a number of benefits to those in natural resource fields as well. They help leaders and others to develop a "curiosity about and understanding of the processes of public policy," a "capacity to assess the causes of conflict and the sources of leverage upon them," "the conceptual and communication skills to use leverage in an effective way," and "an ability to maintain the integrity and vitality of a professional stance while active in often severe contests of values."[34] The case method makes participants aware of the seemingly minor factors that have a major impact on real-life situations. It also focuses on the problem at hand comprehensively, whatever its social and decision context, and re-creates a sense of the immediacy and gamesmanship involved in actual decision making.[35] This method also provides a "stable frame of reference" useful to compare cases over time.[36] Because it grounds understanding of a particular problem conceptually and practically, a case study is one way to structure and communicate understanding of a problem and what to do about it.

Leaders can ask basic questions about each case as a way of drawing lessons from it. In the introductory stages: Was the policy problem stated clearly and simply? Were the purposes of the case stated clearly and simply? Was the point of view of the person telling or writing up the case sufficiently clear? Were case analytic methods clear? Concerning the problem at the center of the case: Was the problem's context (that is, the social process) adequately detailed? Was the problem's status relative to the decision process clear? Were the goals that were sought in reference to the problem clarified? Concerning the analysis of the problem: Were relevant trends (history) adequately described? Were conditions (factors) that shaped trends adequately described? Were future trends (projections) adequately described? Concerning recommendations to solve the problem: Were alternatives to resolve the problem adequately described? Were alternatives adequately evaluated? Was the selected alternative (strategy) or complex of strategies appropriate to achieve goals and solve the problem?

A few case studies from the Yellowstone region have considered these questions.[37] Frost's analysis of the wildlife disease arena, mentioned above, provided useful lessons for active learning. Another case that is full of insights is that of Sue Consolo Murphy and Beth Kaeding (both with Yellowstone National Park), which sought lessons from twenty-five years of controversy over grizzly bear management in the Fishing Bridge area.[38] They were able to show that many of the perils in decision making listed above had occurred, and they identified some ways to avoid them in the future.

These and other good case studies focus on the decision-making process in a contextual, multimethod way; they are based on abundant data; and they seek lessons for future improvements in decision making. Some of the lessons that come from comparative cases focus on the benefits of "civic deliberations" on community- and place-based issues. In all of these, of course, existing environmental laws and regulations must be followed, enforced, and implemented, the expertise of government must still be included, and scientific and technical information is essential and must remain central to civic deliberations. This information can help people understand the nature of the problems and the trade-offs of different choices of solutions. Laws provide the sideboards and the value direction (e.g., Endangered Species Act); deliberative exercises can adapt these to specific situations and integrate diverse values in the process. Agencies retain authority and control and agency leaders have "command of the public stage,"[39] yet civic deliberation requires that agency experts be kept "on tap, not on top."[40] Such deliberations put the agencies and their leaders in a new role to provide information and educate participants in the process. Agency leaders then "serve as champions of the technical merits, commenting publicly when a particular policy proposal . . . is feasible, overly optimistic, or too costly." They should provide a scrupulous overview.[41]

Double-Loop Learning

Active learning demands useful information flows, according to Ronald Westrum, an organizational theorist.[42] Learning is about changing behavior, improving practices over time, and increasing intelligence and effectiveness.[43] Learning is a basic feedback loop—we make decisions based on our goals, take action, and see what happens. The feedback we experience tells us about the results of our actions. In practice we may run through this process only once, a kind of trial-and-error process—this is single-loop learning. In double-loop learning, in contrast, people change their basic operating assumptions when feedback shows that their decisions and actions were inappropriate. Few people target their own assumptions for change as a basis for making better decisions and taking better actions. Double-loop learning goes well beyond the single-loop variety and opens up new opportunities to do better in all ways in the future.

Unfortunately, there are many blocks to active learning, both externally imposed and self-imposed. Lloyd Etheredge, a systems analyst and political scientist, says blockage in government learning comes from several sources.[44] One is the adoption of similar policies across all cases whether they are really similar or not. Another is the repetition of self-blocking behavior (e.g., the "in-the-box" problem). A third is the repetition of a common syndrome of errors in judgment and perception, stemming from being less than fully problem oriented and contextual. There are other blockages too. Etheredge notes that standard theories, such as "failed analytic brilliance, poor design of bureaucracies, inadvertent flaws in decision-making process, or simple cognitive errors" do not explain the record of failed government programs that he studied.[45] He argues that the real blocks to learning are the self-blocks buried in the perspectives of leaders and in the routines of the bureaucracies.

This self-blocked syndrome leads individuals and groups to learn the wrong things from experience. Learning, on the most practical level, is paying attention to what works and what does not work and finding ways to do more of what works and less of what does not. Sophisticated appraisals are needed to make these distinctions, especially in complex policy settings. Good appraisals harvest lessons in a timely fashion and permit us to turn the lessons of hindsight more efficiently into new practices of foresight, in other words, double-loop learning. Leaders and others have many opportunities before them to take the initiative and learn in double-loop ways in the Yellowstone region. In all cases, these opportunities require looking at operating assumptions first and foremost. For many people, this will require a significant departure from their usual modus operandi. It can be done though, as the record shows, and with surprising beneficial outcomes for all involved.

Recruiting and Training Leaders

Greater Yellowstone needs leadership that transcends the immediate, ordinary urgencies. There is so much at stake in the region, and it needs the attention of skilled, smart leaders committed to promoting the public interest. Many people promote the self-regulation of free market economies, but Adam Smith's "invisible hand" is really many hands, each working for private gain in open markets, and the future of greater Yellowstone should not be trusted to such hands. We depend on leaders from all sectors instead to make the crucial distinctions between public and private and between short-term and long-term. To calculate the long-term interest—the common good—requires leaders who carry out an informed dialogue about management policy. There must be a widely shared "vision, knowledge, and a common moral aim" for a policy system to work well, according to Lasswell.[46] Leaders in greater Yellowstone can develop this kind of leadership, recruit and train upcoming leaders, and learn the mix of leadership activities necessary to manage the region toward sustainability in the common interest.[47]

As described in earlier chapters, the skill set needed by leaders and groups who are responsible for developing integrated natural resource management includes critical thinking, keen observation, deft management, and technical mastery. These skills produce leaders who are personally and professionally secure.[48] This type of leader, however, does not spring into being by accident, so serious consideration must be given to developing and promoting to positions of responsibility the kinds of people who are needed. At present it appears that the recruitment process in the Yellowstone region selects for sound administrators rather than transformational leaders. Leadership development is a long-term process that involves several stages of advancement over the span of professional careers. Recruitment relies on the quality of available professional training and perceived opportunities to satisfy personal goals. Advancement typically involves motivating individuals to engage in leadership and politics, find mentors and supporters, build a power base for operations, and successfully pass innumerable screening and selection hurdles. Every stage is shaped by constantly changing variables that are so deeply embedded in bureaucracies that they are largely not open to direct improvement or action. Because of this, direct efforts to shape leadership are typically met with resistance. The best way to find and advance the kinds of leaders who are needed is to target people who are already in leadership positions as well as those coming up who are in midcareer positions, such as program managers, team leaders, and project organizers. Focusing improvements on active professionals who are already in positions of importance saves time and leads to quicker returns because such people have already passed tests of practical politics.

In open, relatively flexible societies we need not only more leaders, but also we need them to be capable of winning support for their policies. The most direct way to get this kind of effective democratic leadership is through a new kind of educational experience. At present no such program exists to train leaders to gain the full range of skills needed. Currently, most education teaches the four-part skill set introduced in chapter 1—critical thinking, observation, management, and mastery of technical subject—only as an accidental by-product of training for science, management, or law. Several options exist to address this shortfall, however, some of which could be organized and carried out in greater Yellowstone, such as short courses and workshops, whereas others would need to be done elsewhere, at, for example, major universities.

Intensive short courses and workshops of a few days or weeks can be designed to improve leadership performance. This option can be designed for agency leaders and those in the nongovernmental world, specifically bright, technically and socially skilled individuals who occupy junior positions yet show promise for advancement. Opportunities must be created for such people to leave their jobs for the time required. The goal should be to improve leadership capacities and qualifications across the whole range of activities discussed in this book. Short courses can offer directed preparation for decision and policy making, building coalitions, conflict management, interpersonal relations, and management of organizational and policy systems. These options should also focus on intellectual capacities, explicit and implicit knowledge, personality, and values. Some workshops of this nature have been conducted in the Yellowstone area.[49] More are needed. Courses of several months or longer would also be helpful in leadership development. A four-to-twelve-month course for leaders both in and out of government could be designed to address the lack of academic background of many participants, the need for participants to unlearn routines, and the need to influence capacities and qualifications. Teaching methods and materials can be adjusted to the unique objectives and participants of each course. Natural resource managers can learn a lot from modern courses designed for government administrators, business executives, and planners. Both passive and active teaching methods, including conventional lectures, guided reading, case studies, and individual and group reports, can be used. For example, complex projects, political gaming, role-playing, sensitivity training, and individual counseling would all be valuable.

To reach the objective of training superior leaders, the courses would have to break through deeply rooted assumptions, opinions, perceptions, and dogmas of the participants, who may well be on the defensive initially. Course organization and location are, of course, also important. Such courses should be systematic and their lessons reinforced adequately over time in order to make a significant difference. The "culture" of these courses would best include intense cooperative effort and high intellectual tension, with hard give-and-take between instructors and participants and among the participants themselves. This requires top-notch faculty, guest lecturers, and a high-quality core staff.

There are many possible programs to convey new leadership styles and problem-solving methods for policy making in greater Yellowstone, all of which could reinforce one another. Individually or all together, these options encourage users to be more problem oriented, contextual, and multimethod in their work.

Adaptive Governance

Leaders need a guide to transition toward sustainability, an approach that will address all levels of the overall problem. Throughout this book I have suggested that transboundary management, which goes by many other names—ecosystem management, watershed management, community-based problem solving—can serve as that guiding principle. Each of these takes flexible forms in different contexts. There is no single, simple formula.[50] The label is far less important than the innovation being called for, and by whatever name, this is what leaders must learn to do well.

The clearest and most complete description of this new way of thinking is what Ron Brunner and Toddi Steelman, who are interested in finding and securing the common interest in managing the environment, call "adaptive governance."[51] It is perhaps our best and most practical means to find and advance the common interest. Because it encourages people to form broader perspectives, it is a potential means of adapting both science and the agencies to the new realities of our rapidly changing world. It opens up new options for leaders and creates a shared sense of opportunity for all people.

Adaptive governance calls for changes in how science, rationality, social process, and decision making are carried out. Brunner and Steelman suggest specific changes in each of these four areas. They argue that science, despite our heavy reliance on it, has been unable to resolve deep value differences in society about conflicting uses of natural resources. Thus, science needs to change from its present usage as a discipline-based science for research on closed systems to an integrative science for practice in open systems; from reductionism to a configurative approach; from prediction, a search for universal laws and mastery of nature through science and technology, to a philosophic commitment to *freedom through insight,* wherein human dignity for all is paramount. These authors call for a science that embraces qualitative, context-sensitive methods that integrate independent data sources into a picture of the whole. Finally, this new kind of science must focus on the behavior of organisms, including humans, which, like all organisms, tend to operate in ways that they perceive (often shortsightedly or incorrectly) will leave them better off.

Rationality, too, needs to be transformed to become our chief means of progress toward active, double-loop learning. Rather than resting on confidence in scientific management and technology and normative models, rationality must become an integrative science, a "policy science," focused on practice and relying on behavioral models. The dominant mode of rationality views goals as single targets that are fixed, given, or assumed, but it must come to comprehend goals as multiples that must be integrated or balanced. It needs to expand from overreliance on quantification to embrace qualitative trends and discrete, unique events. It must explain behavior in terms of contingent multiple factors and take them into account in policy making. It must make projections that are more limited yet open to unknowns and surprises, rather than deducing future events from pertinent "laws." Finally, rationality must scale up resources and methods to match larger geographic areas and scope of coverage.

Brunner and Steelman also call for transforming social process interactions from isolated, discrete events to more interconnected, interdependent, and continuous forms. Rather than being dominated by experts and authorities, the social process must include nonexperts and nonofficials. It must move from a technocratic perspective to one that is democratic and uses the best science. Instead of a bureaucratic, centralized, top-down structure, human dignity would be better served by a relatively informal, decentralized, and bottom-up configuration. From valuing mostly scientific knowledge, technical skills, power, and respect, it must come to include lay or local knowledge and other skills and values in a fair way. From competition in the marketplace of ideas and organizations, it must move to a genuine integration of knowledge.

Finally, Brunner and Steelman call for transforming the decision-making process. We need more sensitive and timely initiation, more factual and comprehensive science and debate. Rather than having plans decided only by authorities, it would be better to get more people involved, in more flexible roles, in planning and appraisal. We need to move from agencies routinely implementing programs to more flexible, contextually sensitive implementation. Appraisal ought to be independent, ongoing, and designed to distinguish successes and failures and to learn actively from them. Termination processes should determine what programs need to be ended in order to free up resources to support alternate, more successful ones.

Ultimately, we need to move from a situation that stifles new practices and institutions to one that encourages them. Adaptive governance is actually a form of integrative politics that promotes common interest outcomes wherever possible. It relies on persuasive strategies, where the force of a better argument wins out over rigid, positivistic science and agency bureaucracy or the blatant exercise of self-interested power. Brunner and Steelman note that the adaptive approach will win out only to the extent that people realize that they are interdependent and must accommodate the interests of others if they wish to advance their own. This transition is now taking place throughout the world in the form of practice-based experimentation, that is, through prototyping and learning on a case-by-case basis. Harvesting the lessons of real-life experience can clarify best practices and open up new opportunities for improvements.[52] In turn, these can be used in the next iteration of experimentation in an endless process of learning.

Conclusion

The single most important concept that can help leaders is the recognition that improving resource management policy fundamentally means improving decision-making processes through the application of the essential tools of active learning, by adaptive leading and managing, and by helping everyone get ahead, in other words, enabling outcomes based on the common interest. "The degree to which any particular individual . . . can achieve his demanded values is a function of the degree to which other individuals . . . can secure and maintain a corresponding enjoyment of their demanded values," notes Harold Lasswell.[53] What this means for greater Yellowstone and its leadership is finding and using an integrated strategy for cooperative problem solving that seeks common interest outcomes.

8 Transitioning Toward Sustainability

It is clear that greater Yellowstone's natural resource challenges involve social values, people interacting, and decision making. Addressing the interrelated challenges—ordinary, governance, and constitutive—demands healthy and functioning institutions, the value-producing and -enjoying structures and processes in our society and skilled, knowledgeable leaders. To move toward a sustainable society and environment, both leaders and the public must come to understand better the value dynamics behind the many conflicts in the region over resources. The conflicts really reflect differences in personal and group perspectives and policies that transcend the ability of scientific management and bureaucratic agencies to understand fully, much less address effectively.[1] In striving to resolve these conflicts and balance people's differing interests, leaders need to guide our social interactions through many bureaucratic, technical, and political hurdles. The stronger the personal commitment and the greater the skill and knowledge of leaders, the greater will be the benefits to the public. John Dryzek, a political scientist, posits that sustainability, concerning as it does "the continuing integrity of the ecological system on which human life depends could perhaps be a generalizable interest par excellence."[2] The sustainability of greater Yellowstone is a test of this notion. We are not alone in our challenges or our struggle to bring forth sustainability, although we may be one of the few places with a strong chance of getting it right. Even little improvements in any of these can make a big difference.[3]

The Current Situation

This book has distinguished the challenges functionally as ordinary, governance, and constitutive. Among benefits of this problem definition is that, by focusing on society's value dynamics and its institutions, it helps us better appreciate the true nature of the challenges, their contexts, and the practical options open to us. It thus offers us a way to escape the many conventional traps in problem solving that are so prevalent today.

We Are Not Alone

A functional viewpoint puts us in a position to see that the sustainability challenges in greater Yellowstone are similar to large-scale resource management problems elsewhere in the world. Although the Yellowstone region is unique biophysically, it shares challenges and possible solutions with other large-scale initiatives worldwide. Stephen Dovers, an Australian natural resource policy specialist, noted that all sustainability problems share universal underlying attributes.[4] He identified several features that can help us put greater Yellowstone in a realistic, global perspective and clarify what needs to be done and how.

The first is that large-scale sustainability problems involve larger spatial and temporal scales than we previously thought.[5] We have enlarged our thinking to include the entire ecosystem beyond the namesake park. We conceptualize the Yellowstone ecosystem as part of an even larger Yellowstone-to-Yukon bioregion, for example. The federal agencies, too, are expanding the scope of their decision making by classifying the United States into different large ecosystem types. Plans are being drawn up in some areas to manage entire regional ecosystems and landscapes, including the Colorado Rivers System, Northern Continental Divide Ecosystem, and the Southern Rocky Mountains. This encouraging trend toward larger spatial and temporal thinking, along with growing interest in integration, is expected to continue.

Second, as part of this new thinking, we are seeing the real possibility of ecological limits to human activities.[6] Our populations and our practices are pushing many species to the edge of extinction and interfering with ecological processes such as pollination, predator-prey dynamics, and forest regeneration patterns—all vital to our own continued existence on this planet.[7] We are disrupting historic migration corridors, fragmenting landscapes, and polluting water and air.

The consequences of these activities, known for a long time, are predictably harmful. We need to appreciate our limits and reverse this trend in greater Yellowstone.

Third, many cumulative, collective actions are having irreversible impacts on the environment and on our society, and measuring these cumulative impacts and determining thresholds of irreversibility are proving difficult.[8] Examples of the connectivity and complexity of such problems are the harm caused by increasing populations, developments, and recreation on and around public lands to sensitive, threatened, and endangered species; oil and gas development affecting air and water quality in far distant mountains and river basins; and disruptions to streams and water flows damaging fishing, boating, and irrigation far down river. The agencies have come to appreciate this fact and have constructed cumulative effects models such as the one designed to measure human impact on grizzly bears. Although this particular model has not been without its critics, it is a step in the right direction in promoting consideration of cumulative effects (and not just limited to technical, quantitative computer modeling). This trend toward cumulative effects and systems-wide thinking needs to be encouraged.

We will never have the precision that would be ideal for managing a large-scale region like Yellowstone.[9] There are limits to the quantification sought by positivistic science, and most of the challenges in greater Yellowstone show inherent associated risk and uncertainty that are not amenable to precise quantification and probabilistic analysis. Bison, elk, and brucellosis science and management; chronic wasting disease and similar problems; the decommissioning of elk feed grounds; and maintaining large mammal migrations—none of these can be studied to the degree that will enable us to understand them with the precision once thought possible. There are no uncontested research methods, policy instruments, or management approaches, yet sound policy judgments and management are called for in the face of risk and uncertainty. This trend toward better appreciation of problems and risks and the need to move forward with informed decisions in the face of uncertainty should also be promoted.

We also cannot reduce natural resources and their management to monetary values, despite the efforts of many people to do so. It is not even possible in many cases to assign monetary values to ecological services, for example. What are predator-prey relations worth? Or personal experiences in the wilds of Yellowstone? Or solitude and quiet? Trying to "commercialize" or "marketize" greater Yellowstone and all its human and natural transactions merely perpetuates the old ways of doing business that caused many of today's problems, and we certainly do not want to turn public policy making over to the free market with all its distortions, externalities, and foibles. Greater Yellowstone's problems cannot be reduced to a calculus by which we can determine public and private costs and benefits of our present resource uses, much less the outcomes of future policy interventions. Both natural and human systems are too complex for that. We must all resist this trend to marketize our natural resources.

We now understand that the basic cause of most, if not all, of the present resource management problems is the way we have organized our society, our system of values and institutions, and our underlying beliefs about the position and role of humans in relation to nature. Moving society toward more sustainable practices means shifting some of those basic organizational variables, especially constitutive and governance processes, and responding to growing demands for more community participation in management policy. Demands for greater participation are typically resisted or frustrated by existing government agencies, special interests, and political elites. The agencies are particularly adept at blocking legitimate democratic demands.[10] We must open up problem solving and decision making to be more participative, responsible, and accountable. These are weighty matters that are worth serious consideration and should be at the top of the public agenda for deliberation.

Leadership Skills and Judgment

There are many experienced leaders within and outside government in greater Yellowstone who have been working to achieve a healthy and secure future for us all. In order to become more effective they must better appreciate the interdependencies and interrelationships among natural resources, interests, and institutions, all of which affect value processes in the region and beyond. But many still think and act on a unit-by-unit, year-by-year, agency-by-agency, and interest-by-interest basis. The patterns of usage and control of greater Yellowstone's sharable resources have become increasingly important and increasingly controversial. This trend may be expected to continue, but these cascading challenges will also open new possibilities and options to improve matters.[11] Knowledgeable, skilled leaders can and must address these problems and processes effectively.[12]

The current problems of leadership stem from the level of knowledge and skill that individual leaders have brought to bear on these complex problems. Most leaders in the agencies, for example, were hired to administer the national parks, forests, and wildlife refuges, not to coordinate and integrate across all the federal agencies for sustainability in a new interdisciplinary and democratic way. They are being asked to address new challenges using organizational means that were designed, structured, and operated to address relatively routine, conventional, ordinary problems through scientific management principles.[13] But these principles are outmoded; they do not fit our more experienced understanding of the nature of the problems. But taken all together, the problems facing leaders are not merely a collection of ordinary problems that can be resolved by carrying on business as usual. They are, in fact, a product of the institutional system's inherent flaws. Conventional understanding, traditional bureaucratic responses, and standard-issue administrative leadership styles cannot solve these evolving challenges to our systems.

Among the skills leaders can cultivate to move us toward sustainability is avoiding the many conventional traps inherent in traditional technical disciplines; in bureaucracy; and in the advocacy, litigation, and education activities of many NGOs. One such trap is the "marketization" of greater Yellowstone. The rush to marketize the arena comes from the failure of past technical, scientific, and economic "solutions" to problems, which have led to inadequate policy responses to the real problems, according to Mark McBeth and Elizabeth Shanahan, two political scientists from the region.[14] At present, many groups and communities in greater Yellowstone are acting as policy marketers who market public opinion to citizens and to each other, an approach that mirrors the rise of consumerism in our culture. What all this marketing does is lead to more conflict and a compounding of the governance and constitutive challenges. It does not lead to deliberative, well-grounded policy responses to the actual challenges. Another potential trap for leaders is the pathological power-balancing strategies and bureaucratic politics that currently dominate decision making. Officials and citizens who want to break out of that dynamic find it very difficult to change the agencies or the politics within which they work.

We need to find ways to help leaders and society overall avoid these common traps. We need to set up new ways for constructive change to take place. Nongovernmental leaders have an especially important role in this because of their relatively greater flexibility as compared to governmental leaders. According to policy scholar Sheila Jasanoff, NGO actions in building or rebuilding the knowledge-to-action connection take three major forms, each of which offers opportunities for collaboration and integration.[15] They are criticism/reframing, the building of epistemic communities (for example, informal groups for producing knowledge and bridging conceptual and political divides), and technology transfer. Leaders in NGOs in greater Yellowstone can explore these modes of reconnecting knowledge and action so they can be effective in bringing about the transition toward sustainability.

All of the present leaders have key roles to play in any future transition. They can enable the political and institutional frameworks that will allow people to work together to clarify, secure, and sustain their common interest. They can show a genuine willingness to experiment and assume risk when working with professionals outside their organizations, in academia, and the citizenry. Finally, they can demonstrate a shift away from the dominant, but dysfunctional, scientific management worldview to one that integrates different knowledge sources and frameworks.

Accentuating Positive Trends

A long-term overview of society shows that people's views about their relationship to the environment have changed dramatically in the last half century. This trend is evident in the work being done in greater Yellowstone. Some changes in greater Yellowstone are moving us closer to sustainability, and we need to think more rigorously and practically about accelerating these positive trends. A successful transition toward sustainability will require much more thoughtful, on-the-ground action than we have seen so far in the region. Leaders and the public will have to become more contextual, strategic, and action oriented.

Shifting the Focus of Attention

The trends and conditions moving us closer to sustainability are evident in recent history. In the 1960s and 1970s Americans became increasingly alarmed about environmental problems, such as deforestation, species extinctions, desertification, soil depletion, loss of fish stocks, wildlife habitat, toxic poisons and pollution, ozone loss, and buildup of greenhouse gases. We realized that we were drastically damaging the earth's life support system and reducing our own health, our prospects for economic well-being, and the likelihood of a good life for future generations. These issues were joined on the public agenda by a number of social problems, including rapid human population growth, poverty, growing illiteracy, and social injustice. Our solution at the time was to turn to better technology (such as pollution reduction) and new laws (such as the Endangered Species Act).

Today, although many Americans still have a strong faith in technological and legal fixes, there is a growing sense that broader solutions are necessary. It is now widely recognized that better technology and legislation by themselves will not give us the kind of future we want. They will not reverse or redress the myriad ordinary, governance, and constitutive problems in greater Yellowstone. We now accept that environmental damage is rooted in our fundamental beliefs and the values we hold about modern society and our relation with nature. As Lynton Caldwell put it decades ago, "The environmental crisis is an outward manifestation of a crisis of mind and spirit. There could be no greater misconception of its meaning than to believe it to be concerned only with its symptoms such as endangered wildlife, man-made ugliness, and pollution. These are all part of it, but more importantly, the crisis is concerned with the kind of creature man is and what we must become in order to survive."[16]

Decades ago we began to shift our thinking, language, and actions worldwide toward a sustainable society. No one has defined this notion precisely, and there remains vigorous discussion about what it means. The landmark 1987 Gro Brundtland report from the World Commission on Environment and Development used the term *sustainable development,* calling for development that meets "the needs of the present without compromising the ability of future generations to meet their own needs."[17] Today, it remains a powerful idea. There is consensus that foresighted change is needed in managing human affairs. Moreover, there is broad agreement that we must change the way we think, what we value, and how we behave if we are to have a healthy future.[18] We face a huge relearning challenge, forced on us by the clear lack of sustainability in our institutions and management policy.[19] We must find ways to maintain the functioning of the planet's ecosystems as well as our societies and institutions.

Sustainability, as a concept and as a set of practices, is the culmination of decades of societal learning. Policy scientists Natalia Mirovitskaya and William Ascher define a sustainable society as one "that is based on a long-term vision in that it must foresee the consequences of its diverse activities to ensure that they do not break the cycles of renewal; it has to be a society of conservation and generational concern. It must avoid the adoption of mutually irreconcilable objectives. Equally it must be a society of social justice because great disparities of wealth and privileges will breed destructive disharmony."[20] In pursuing these goals we must be careful that the concept and the term *sustainability* not become a slogan useful only in sound bites. It must be given meaning with strong theoretical and practical grounding.

It must become both a rallying call and a set of practices for integrating environmental and human concerns.

The lessons are clear. We need to learn to be sustainable everywhere, but each place requires a different application because the context is different. The fledgling effort toward sustainability in greater Yellowstone is just one among many worldwide. In greater Yellowstone some problems are small, but others are large. In the face of this range of small, short-term problems to big, long-term problems, most current effort is focused on coping with the small ones. The sheer volume of these small problems prevents transformative-tending leaders from having the time to think and reflect, much less address higher-order tasks. It precludes them from considering the full range of challenges and consequences in depth. Virtually all management policy issues are highly political, and they are getting more so over time. Fear of politics keeps leaders focused on the here and now. These forces and factors conspire to constrain the creativity and scope of our leaders. To get out of this pattern, they need to shift their focus of attention and work toward higher-order tasks. This is possible but will occur only if this new kind of leader is actively supported.

Leaders and Change

Many of our present leaders in greater Yellowstone are rich in experience but poor in theory—about both the challenges they and we face and the responses needed. As a result, opportunities go unrealized and leaders often do not know what to do except what they have done in the past. Part of the fundamental problem that we all, leaders and citizens alike, face is coming to grips with the nature and degree of change that is taking place in the human community, its effects on the environment, and how that change can be used to make improvements toward sustainability. Today, unprecedented change is occurring in all value processes and associated institutional practices. We must try to understand change and use it to advantage.

Among the key features of change in our times, according to Yehezkel Dror, the policy scholar, is that a lot of change is rapid and nonlinear.[21] Not only is it coming fast, but also it is not a linear extension of past trends. Much change is exponential and therefore hard to predict or appreciate. Also, much of it shows a kind of complexity that is multiplying rapidly and therefore hard to understand; it leads people to intense frustration, trauma, and unrest in the more difficult situations. Dror argues that governance systems everywhere must be responsive to these features of change, but in many cases are not.

Currently, many opportunities exist to advance toward sustainability through successful transboundary management, although this requires overcoming significant existing boundaries—geographic, organizational, special interest, disciplinary, epistemological, and others. The task before the Greater Yellowstone Coordinating Committee and other leaders is to break down and overcome these many boundaries. Without transforming itself, the GYCC cannot lead the transition toward sustainability. Change opens up numerous possibilities for innovations and new practices that can be spread and adopted, if leaders are prepared to act. Transboundary management is an innovation in greater Yellowstone that holds great potential, and what leaders ultimately decide to do with this innovation today has significant consequences for the future of the region. Throughout the world, the idea of ecosystem or transboundary management is gaining currency for three reasons, even though some powerful interests are working hard to restrict it. First, the true costs of traditional resource extraction exceed the benefits, and the statistics of biodiversity attrition and other forms of environmental degradation have reached unacceptable levels in the view of many people. Second, the depth of awareness of these issues and the need for a shift in policy in the general population has increased dramatically through education by the popular media, which in turn have drawn on environmental NGOs and scientific studies. Third, the idea of regionwide transboundary management evokes powerful symbols—a sustainable healthy future; long-term economic, aesthetic, and ecological viability; citizen empowerment; and enhanced democratic participation in government decision making, to mention only a few. The time is right for leaders in greater Yellowstone to throw their weight fully behind the innovation of transboundary management.

Change and innovation have a mixed record in Yellowstone for several reasons. For one thing, ideas like transboundary management seldom translate neatly into crisp official responses. Also, elected and agency officials, who are key leaders in the region, live in a policy-making world that is fragmented, consisting of semi-autonomous state and federal bureaucracies and many diverse competing interests, and the structures of bureaucracy clearly get in the way of resolving differences. Any genuine transition toward sustainability requires changes not only in natural resource management itself, but also in the very nature of the professional and institutional systems that manage the region.

The idea of a new policy for transboundary management is in itself a powerful idea.[22] First, it has changed the way people talk about the region, its management, and the federal agencies that play such a pivotal role there. The GYCC's very existence and its goals speak to the power of this idea. The message of this new language is clear. It lays out an attractive vision for the future and erodes the legitimacy of old, traditional management and the interests that support it. Second, it has changed the allocations of responsibility, of blame and credit, for the costs and benefits of traditional management by the agencies. This shift in responsibility is obvious in the growing defensiveness of some interest groups, such as the "wise use movement" and some agencies. It is also obvious in the growing activism of many NGOs in promoting ecosystem and transboundary policies. Third, it has changed awareness among citizens, businesses, and politicians concerned with the Yellowstone region and as a consequence changed the patterns of citizen mobilization. The idea has aroused concern and attracted attention by linking management policy problems to values. In turn, values and ideas of nature (such as ecological resiliency) and human sustainability (including dignity for all) stimulate and channel activism both within and outside the agencies to influence government. The idea of a greater Yellowstone management policy eats away at the power of traditional commodity extraction interests and old, conventional management philosophies deeply rooted in agency cultures and SOPs, in traditional business arrangements, and in antiquated policy systems. Leaders can capitalize actively on these powerful trends to move us forward.

Greater Yellowstone's Future

We are in the early days of a transformation toward sustainability, which is taking place much more at the individual level than at the organizational, institutional, or policy levels. What is changing is the way individuals think about and conduct science, management, and policy, how they think about their organizations and institutions, and how they make public and private decisions. For example, people are increasingly knowledgeable, attentive, and sophisticated about the environment now. We are slowly beginning to shift our own values and our society toward sustainability. More and more people are coming to terms with an integrative style of thinking and operating, bringing them closer to our goal.[23]

Several alternatives suggest themselves to the challenges that we face in greater Yellowstone. The first possibility is simply to stay with what we know best, the status quo, which appears safe, at least in the short run. But as policy scientists Ron Brunner and Toddi Steelman have noted, "Business as usual in natural resources policy at the national level or local level is probably not sustainable over the long run, even if it is adapted to the short-run narrow interests of participants."[24] In short, exploitive single- and multiple-use management, the type that many individuals and agencies carry out today, have caused the problems we face now. More of the same behavior will not solve the problems. Besides, there is urgency in today's environmental problems,[25] and as one policy specialist has pointed out, "Facts and values are always closely intertwined and there never is a starting point."[26] Our problems cannot be successfully addressed by "pooling more and more additive information to already outdated cores of knowledge."[27] We do not need an expansion of existing scientific management, knowledge pieces, and responses added onto the old network of explanations and ways of doing things. The status quo will not bring about sustainability.

The second option calls for somehow taking a quantum leap forward or starting over, beginning with a clean slate to produce much better management policy than we have at present. Practically, such drastic change is not realistic. Radical change, leading to an improved "design rationality," would entail people moving outside and beyond their original frames of reference and fundamentally redefining the challenges.[28] Creating such a reflective and discursive policy process would resolve otherwise intractable policy dilemmas and controversies, but it would require so significant a leap for individuals, institutions, and society from current approaches that it is extremely unlikely in greater Yellowstone today or in any other setting. Regardless of the need, a total housecleaning is unlikely to occur spontaneously, and if it did, it would likely be accompanied by major cultural trauma.

The third, more sensible option is the middle way, transitioning toward sustainability by improving existing management policy, changing problem-solving strategies, and transforming leaders' skills and behavior. This requires imagination, creativity, the upgrading of skills, and the development of a practical response that permits leaders and citizens to orient to problems comprehensively and practically.[29] This would give us the capacity to move to the integrative understanding and action that are needed.[30] For such a transition to happen in greater Yellowstone or anywhere else, Lun-Chu Chen, a law professor and promoter of human rights and international justice, says that leaders (and individual citizens) should think and act *globally* (to meet the growing challenge of global interdependence and interdetermination among all people and countries). They should also act *temporally* (about the present but also about future generations). And they should act *contextually* (so that all decision making relates to all community levels and all value sectors of social interaction). Finally, they should act *creatively* (to mobilize all available problem-solving skills and resources).[31] Taking these actions, he argues, will promote human dignity and sustainability.[32] His advice can help animate leaders and citizens alike in greater Yellowstone to invent sustainable management policy.

The urgency of problems calls for the creation of efficient interaction between natural and social scientists and the public through policy-oriented, integrative concepts and tools. It also requires a reconceptualization of the policy process away from scientific management and bureaucracy. This demand for interdisciplinary research, new social mechanisms, and new institutions is increasingly seen as essential to efforts to make our communities and country more sustainable. The growing governance and constitutive crisis can be avoided if problem solving and social and decision processes benefit from responsible, knowledgeable, and skilled leadership.[33]

Yellowstone is a global treasure, however one thinks about it as a place, a park, region, or ecosystem. Today, the legacy of Yellowstone in these many incarnations has come down to us to cherish, use, and pass on to future generations. There has always been public debate about what Yellowstone is, what it means, and how we should use it. Presently, the debate is heating up as competing interests claim what they see as their share of the Yellowstone ecosystem for themselves, often at the expense of other legitimate interests. Yellowstone belongs to us all, and our leaders are in a unique position to help us secure it sustainably for present and future generations. Greater Yellowstone faces numerous problems as a consequence of past uses and the present social and institutional system. Its leaders are grappling with these interactive challenges in diverse ways, focusing mostly on coping with numerous ordinary problems. The cumbersome bureaucracy and rapidly changing context make addressing all problems that much more difficult.

In the United States and elsewhere, natural resources and their management are becoming important and increasingly controversial components of public discourse. As the Yellowstone region and the world become more interdependent, as natural resource availability becomes more limited, and as people everywhere become more conscious of the status of the environment and the precariousness of the human enterprise, new demands are being made on leaders to upgrade resource management policy toward sustainability to serve the common interest. Citizens, ecologists, policy analysts, and policy makers increasingly recognize the paramount importance of the interdependencies and interrelationships among natural resources and between resources and the societal institutions that exploit them. What we need now is to turn this growing awareness into better accomplishments at all levels in natural resource management policy.

The way forward is clear. We will need to shift the focus of attention, accentuate positive trends, and improve democratic governance and constitutive processes. These require raising the consciousness of people, where needed, about the challenges and opportunities they face. They require recasting how problems are understood and shifting to problem definitions and solutions that are more realistic, comprehensive, and practical. They require encouraging new patterns of human interaction that open up opportunities to address problems more cooperatively and effectively, that is, more democratically. If knowledgeable and skilled leaders and citizens meet these three needs, we can achieve sustainable natural resource use in greater Yellowstone.

Appendix 1
Official Goals of the Greater Yellowstone Coordinating Committee

Goals and goal-like statements of the Greater Yellowstone Coordinating Committee (GYCC), beyond those listed in the text, are summarized in this appendix. For example, in April 1994 the minutes of the group's meeting listed six goals.[1] Under a section labeled "Purpose and Role," the minutes noted: "The Yellowstone ecosystem . . . is a unique and special place. Lands in the Greater Yellowstone Area (GYA), administered by six National Forests and two National Parks are geographically contiguous, ecologically interdependent and unalterably linked. Unit boundaries are largely invisible on the ground and can be difficult for the public to appreciate." They go on to state, "members of the Greater Yellowstone Coordinating Committee (GYCC) recognize that our ethical duty is to cooperatively manage Greater Yellowstone Area (GYA) resources, sustaining existing values and characteristics, consistent with the missions of the Agencies." Finally, they set these goals for the GYCC:

1. Provides leadership in making decisions that serve the public and manage GYA resources.
2. Sets priorities and assigns resources to achieve objectives.
3. Coordinates planning across units, at all levels, with due consideration of overall needs of the GYA.
4. Ensures coordination of strategies and practices across National Parks and National Forest Service units.
5. Minimizes duplication of effort.
6. To the extent possible by law and agency missions, makes rules and regulations consistent across GYA. Inconsistencies will be the exception and where they occur, there will be good and logical reasons.

These goals are a streamlined version of the 1985 goals cited in the text, and they do emphasize some of the higher-order tasks required for success—decision making, coordinating, and integrating, rather than taking a case-by-case focus.

In June 2001 the GYCC's newly printed Briefing Guide proclaimed GYCC's motto, "Transcending boundaries in one of America's most treasured ecosystems," at the top of the first page.[2] This slogan captures all of the GYCC's goals in one short statement. In a section entitled, "What is the role of the GYCC?" this same document listed the six 1994 goals (see above). This document also noted that the GYA is "geographically continuous, ecologically interdependent, and unalterably linked."

Other GYCC records show somewhat different goals.[3] The minutes from September 1994, which included a review of the 1986 memorandum of understanding, listed as its goals to: (1) establish an integrated information management process (data management and GIS capability) for GYCC managers that would facilitate public land management throughout the GYA, and (2) develop this integrated information base with the concept of a "dynamic assessment" of social, ecological, and economic factors. In the minutes it was recorded that "dynamic assessment" was the way the GYCC should go. Data acquisition should not be driven by one or more major issues across the GYA. Data would not be gathered on a case-by-case basis. Instead, priorities for data acquisition would be based on the working needs of all the GYCC managers. Assistant Secretaries of Agriculture Jim Lyons and George Frampton were present at this meeting. A historical overview of the GYCC was given by Dave Garber and Jan Lerum to the group and these two secretaries. The official goals from this point seemed to focus on data acquisition and integrating information for use by unit managers. Thus, it is unclear if the fourteen official goals in the 1985 memorandum of understanding and the six goals from earlier in 1994 were to take a back seat to data gathering and building a geographic information system (GIS) mapping capacity.

A report from the April 1996 meeting of the Unit Managers Team (that is, the GYCC) said, "The role of the Unit Managers Team (UMT) is to provide leadership, guidance and coordination for the National Parks and National Forests in the Greater Yellowstone Area.[4] The GYCC's purposes, stated in five goals, were to:

1. Help formulate and support the integration of the *Framework for Coordination of National Parks and National Forests in the GYA* into agency plans and activities.

2. Help set the climate and attitude of meeting public-customer expectations.
3. Share information and resources.
4. Identify and provide for the resolution of emerging issues within the Greater Yellowstone Area.
5. Provide a forum for interaction with other federal, state, and local agencies and private organization and the public.

Membership in the committee, meeting format and agendas, and other elements of the interagency and interunit relationship were laid out in general terms. There were eight signatories to this agreement, including the six national forest supervisors and two national park superintendents. Assistant Secretary of Agriculture George Frampton was in attendance.

A May 1997 meeting that was closed to the public revisited goals. At this meeting the GYCC also reviewed the GYCC charter (1996 Leadership Guidance Coordination in GYA document), the historic overview from 1994 by Dave Garber and Jan Lerum, the GYCC's 1994 purpose and role, and the 1986 memorandum of understanding as background for future actions. This was clearly a major effort to rationalize its goals. Minutes indicate this goal-setting exercise produced several lists.[5] First, roles of the Unit Managers Team were detailed in fifteen entries, among them goals about coordination and cooperation across units and agency boundaries, providing leadership, and monitoring progress. Second, operating principles/guidelines were offered in twelve entries (for example, operate by agreement, think beyond our singular selves, controversial things stay in this room). Third, the format of future meetings was decided (among its nine goals were to be strategic more often than tactical and to look at how best to involve the public). Fourth, a list of thirty-three "specific issues," all in fact operational goals, was generated (for example, winter use assessments, elk winter range and migration routes, large carnivore, public expectations in GYA). Fifth, fifteen opportunities for cooperation were listed in such areas as fisheries, watershed groups, fire management, large mammals, enlarged research, and public information. And sixth, nine future issues were identified, including transportation planning, user fees, large mammal issues, newsletters, and developing GYCC "talking points." When a group produces lists, so many lists, it indicates that it is not spending enough time analytically orienting to the problems and context that it faces.[6] Producing lists is a common conventional substitute for engaging in the kind of systematic, problem-oriented work that is really needed to understand challenges and set goals to address them.

Additionally, the May 1997 meeting highlighted data needs, research needs, and funding opportunities in public information, recreation use, land patterns, waterways/fisheries, air quality, fire management areas. Each of these substantive areas was discussed and a goal was formulated. For public information, the goal was to provide more effective information for the public to choose from, by which the GYCC meant more coordinated and easily usable information. For recreational use, the goal was to establish present and future recreation needs and trends. A task force was assigned, similar to the winter use one, to address this task. For land patterns, the goal was to manage for ecological integrity of the greater Yellowstone for the public benefit. For waterways/fisheries, the goal was to restore and/or conserve native species (with the recognition that habitat degradation is not the sole cause). For air quality, the goal was to get an overall assessment of air quality and problem sources. And for fire management, the goal was to return fire to the ecosystem in a socially acceptable way. Each goal statement was accompanied by an explanation and a statement of data needs, research needs, and funding opportunities. The motivation for this major goal clarification effort was never made clear.

In 2001 the committee restated its official goals.[7] Again it noted that members have a "responsibility to cooperatively manage GYA resources to sustain existing values and characteristics, consistent with the missions of the agencies."[8] Six goals were listed:

1. Provide leadership in making coordinated decisions that serve the public and help sustain the resources. Ensure coordination of planning, strategies and practices across national park, national forest, and national wildlife refuge units.
2. Set GYCC level priorities and assign resources to achieve objectives.
3. Provide a forum for interaction with federal, state, local agencies, private organizations and the public. Help foster a climate that encourages coordination and sharing.
4. Identify and provide for resolution of emerging issues within the GYA.
5. Minimize duplication of effort; seek opportunities to share information, resources, and data.
6. To the extent permissible by law and agency mission, make rules and regulations consistent across the GYA.

In 2003 the committee and the agencies produced another memorandum of understanding. This document listed the purpose to "document the intentions and provide a framework for providing public services and responsible land management in a cooperative and coordinated manner throughout the GYA, to the extent permissible by law and agency missions, throughout the GYA."[9] Under objectives it noted: "It is the desire of all parties to cooperate fully, in manners relating to responsible land management throughout the GYA. These cooperative efforts include, but are not limited to:

1. Provide leadership in making coordinated decisions that serve the public and help sustain the resources.
2. Ensure coordination of planning, strategies, and practices across national park, national forest and national wildlife refuge units in the GYA.
3. Set GYCC level priorities and assign resources to achieve goals.
4. Foster a climate that encourages interaction, coordination and sharing with federal, state, local agencies, private organizations, and the public.
5. Identify and facilitate resolution of emerging issues within the GYA.
6. Minimize duplication of effort; seek opportunities to share information, resources, and data.
7. To the extent permissible by law and agency missions, makes rules and regulations consistent across the GYA."

These objectives are very similar to those from 2001.

Appendix 2
Agenda Topics Listed for Greater Yellowstone Coordinating Committee Meetings from 1995 to 2004

Topic/Issue	1995	1996	1997	1998	1999	2000	2001	2002	2003	2004	Total
Social assessments	√										1
Grizzly bear model/study	√	√		√	√			√	√		6
Information management	√	√									2
Science working group	√	√		√							3
Budget	√										1
New chair	√	√	√		√	√	√				6
Direction, philosophy, expectations, structure, procedures for GYCC		√	√	√		√	√	√	√		7
GYCC priorities		√				√	√	√	√	√	6
Funding		√			√	√		√	√		5
Data collection/ management		√				√					2
Vegetation mapping/ data base/training		√									1
Partnership opportunities		√									1
Winter use management		√		√	√	√		√	√	√	7
Fire management	√	√	√			√		√	√	√	7
NPS fee collection	√										1
Waterways/watershed/ fish (cutthroat trout)			√	√		√	√	√		√	6

Topic/Issue	1995	1996	1997	1998	1999	2000	2001	2002	2003	2004	Total
Rocky Mountain Elk Foundation			√								1
Air quality and transportation			√	√	√	√					4
Heritage Trust				√	√						2
Open space				√							1
Public information				√							1
Recreation use/fees				√	√					√	3
Land patterns				√		√	√				3
Weed management				√				√			2
Land acquisition				√							1
Campground reservation system				√							1
Fund raising				√				√			2
Forest orders on food storage				√							1
Diversity in the workforce				√							1
Commercial use permitting				√							1
Sheep management				√							1
Soils mapping				√	√						2
Backcountry management					√						1
Wilderness/wildlands					√		√	√			3
Community monitoring program					√						1
DNA microbial research					√						1
Systems engineering					√						1
Cellulose conversion					√						1
Interdisciplinary problem solving					√						1
Invasive species						√	√				2
Jackson elk and bison management						√		√			2
Clean Cities Coalition						√					1
GYCC records/archives						√					1
Project review/ solicitation						√	√	√	√		4
Whitebark pine						√	√	√			3
Trumpeter swans						√					1
Wolverines						√					1
Yellowstone Teton Regional System							√				1
Strategic research and science strategy							√				1
Memorandrum of understanding MOU/charter/ amendment								√			1

(*continued*)

Topic/Issue	1995	1996	1997	1998	1999	2000	2001	2002	2003	2004	Total
History of GYCC								√			1
Interagency cooperation—federal highway projects								√			1
Inventory and monitoring								√			1
Human safety/ sanitation								√			1
Forest land management plan									√		1
Wildlife									√	√	2
Private conservation										√	1
Subtotal of new issues addressed	6	9	3	14	7	8	2	5	2	2	
Subtotal of previously addressed issues	0	4	3	6	8	11	9	13	8	4	
Total issues addressed	6	13	6	20	15	19	11	18	10	6	

Note: Agendas from four meetings were missing: January 19–20, 1995; April 25, 1995; April 3–4, 1996; and May 14, 1997.

Appendix 3
Projects and Costs Funded by the Greater Yellowstone Coordinating Committee, 2000 through 2004 (data from GYCC annual reports)

Project Title	Funded June 1, 2000	Funded March 26, 2001	Funded April 15, 2002	Funded March 5, 2003	Funded January 27, 2004
Madison Ranger District backcountry weed management (BDNF)	$10,000	$15,000	$12,000	$12,000	
Gravelly Mountains land exchange (BDNF)	$5,000				
Winter use monitoring (BDNF)		$4,000		$4,000	
Inland West Watershed Initiative strategy		$1,000			
Westslope cutthroat trout broadstock program (BDNF)				$5,000	
Wolverine land exchange (BDNF)			$5,000		
Winter range weed control					$4,000
Gravelly Mts. bear conservation (BDNF)					$7,000
Winter visitor use monitoring (BDNF)					$4,000
Teton integrated noxious weed control (BTNF)	$12,000				
Teton wilderness salt site study (BTNF)	$10,000	$10,000			
Teton division visitor support (BTNF)	$6,000				
Purge spurge (BTNF)					??

(continued)

Project Title	Funded June 1, 2000	Funded March 26, 2001	Funded April 15, 2002	Funded March 5, 2003	Funded January 27, 2004
Resort naturalist program (BTNF)		$5,000			
Winter use monitoring (BTNF)		$6,000		$6,000	$6,000
IWWI watershed strategy (BTNF)		$1,000			
Wolverine ecology and management in the Teton Range, Wyo. (BTNF)			$10,000	$5,000	
Cutthroat trout distribution mapping (BTNF)			$10,000		
Implementation of forestwide food storage/sanitation order (BTNF)			$9,000		
"Share the Trail" collaborative planning process (BTNF)			$10,000		
Snake River and Yellowstone cutthroat trout subspecies distribution mapping (BTNF)				$10,000	$10,000
Hobble Creek restoration (BTNF)				$6,000	
Bridger Wilderness noxious and nonnative weed treatment plan and implementation (BTNF)				$2,000	
Front Line Assistance program for noxious weed control (BTNF)				$5,000	$5,000
Watershed improvement and restoration field inventory crew (BTNF)				$5,000	
Confluence Information Center (BTNF)					$5,000
Salt site compliance (BTNF)					$4,000
Noxious weed inventory and mapping (CTNF)	$10,000				
Role of bears in trout habitat and hydrological function (CTNF)	$4,500				
Pine Creek fish weir (CTNF)	$10,000				
Yellowstone cutthroat trout distribution surveys, South Fork Snake River (CTNF)	$6,000	$10,000			
Whitebark pine planting (CTNF)	$3,000				

Project Title	Funded June 1, 2000	Funded March 26, 2001	Funded April 15, 2002	Funded March 5, 2003	Funded January 27, 2004
Palisades Watershed Area noxious weeds (CTNF)		$8,000			
Black Canyon leafy spurge (CTNF)		$7,500	$5,000	$5,000	
Trumpeter swan nest habitat (CTNF)		$7,500			
Wolverine denning habitat survey (CTNF)		$10,000			
Winter use monitoring (CTNF)		$6,000		$6,000	$6,000
Willow Creek Research Natural Area—weeds (CTNF)			$2,000		
Heli-mapping of leafy spurge (CTNF)			$2,000		
Dubois Ranger District weed mapping and treatment (CTNF)			$2,500		
Wolverine ecology and management in the Teton Range, Wyo. (CTNF)			$7,500	$5,000	
Morris Creek leafy spurge (CTNF)				$5,000	
Highlands "Bag of Woad" (youth weed treatment program) (CTNF)				$5,000	
Montane wetland characterization and management (CTNF)				$3,000	
Thomas Fork Bonneville cutthroat trout passage project (CTNF)				$7,000	
Yellowstone cutthroat trout distribution surveys and mapping on Caribou-Targhee NF (CTNF)				$6,000	
Burns Creek trout passage (CTNF)					$6,500
Sawtell Creek habitat improvement/restoration (CTNF)					$10,000
Fish Creek grazing exclosures (CTNF)					$2,500
Winter recreation education/signing (CTNF)					$5,000
Yellowstone cutthroat interpretive display (CNF)	$5,000				

(continued)

Project Title	Funded June 1, 2000	Funded March 26, 2001	Funded April 15, 2002	Funded March 5, 2003	Funded January 27, 2004
Yellowstone River/ Yellowstone cutthroat trout (CNF/GNF)	$10,000				
Beartooth Highway noxious weeds (CNF)		$10,000			
Beartooth Highway corridor plan (CNF)		$10,000			
Winter use monitoring (CNF)		$2,000		$2,000	$2,000
Rock Creek weed treatment (CNF)			$5,000		
Cutthroat trout distribution mapping (CNF)			$8,000		
Yellowstone cutthroat trout population inventory/ distribution mapping (CNF)				$7,000	
Creation of Carbon and Stillwater County Cooperative Weed Management Area (CNF)				$5,000	
Noxious weed treatment along the Beartooth All American Road corridor (CNF)				$7,500	
Whitebark pine inventory (CNF)					$4,000
West Yellowstone public lands information desk (GNF)	$5,000				
Basin noxious weeds inventory (GNF)	$5,000				
Royal Teton Ranch lands work (GNF)	$10,000				
Impact of backcountry use on large carnivores (GNF)	$5,000				
Whitebark pine planting (GNF)	$5,000				
Yellowstone cutthroat trout video (GNF)	$3,000				
Duck Creek wetlands land work (GNF)		$5,000			
Slip and Slide conservation easement (GNF)		$5,000		$5,000	
OTO Ranch management plan (GNF)		$5,000			

Project Title	Funded June 1, 2000	Funded March 26, 2001	Funded April 15, 2002	Funded March 5, 2003	Funded January 27, 2004
Upper Gallatin invasive species (GNF)		$10,000			
Effects of recreational use on grizzly bears (GNF)		$10,000	$8,000	$3,000	
Upper Yellowstone water management area (GNF)		$5,000			
Gallatin River westslope cutthroat (GNF)		$5,000			
Winter use monitoring (GNF)		$6,000		$6,000	$6,000
Inland West Watershed Initiative strategy (GNF)		$2,000			
Absaroka-Beartooth wilderness and whitebark pine blister rust status (GNF and CNF)			$3,000		
N. Yellowstone winter range-weed mapping/control/modeling (GNF)			$3,000		
Cutthroat trout distribution mapping (GNF)			$10,000		
Implementation of forestwide food storage/sanitation order (GNF)			$10,000		
Yellowstone cutthroat and rainbow trout in the upper Yellowstone river-distribution mapping (GNF)				$7,500	
320 Ranch/Taylor Fork land acquisition and related land use actions (GNF)				$7,500	
Digital baseline map for off-highway vehicle use in West Yellowstone, Mont. (GNF)				$3,000	
Saving Lives through Avalanche Education (GNF)				$3,000	
Whitebark pine status in the Absaroka-Beartooth Wilderness (GNF/CNF)				$4,000	
Beartooth Wildlife Conservation Management Area development (GNF)					$6,000
Weed control in Absaroka-Beartooth Wilderness (GNF)					$3,000
Human use/campsite trends in Absaroka-Beartooth Wilderness (GNF)					$5,000

(*continued*)

Project Title	Funded June 1, 2000	Funded March 26, 2001	Funded April 15, 2002	Funded March 5, 2003	Funded January 27, 2004
Duck Creek land acquisition (GNF)					$10,000
Gros Ventre River spotted knapweed (GTNP/BTNF)	$20,000	$7,500			
Wolverine survey and monitoring (GTNP through CNF)	$10,000				
Effects of irrigation on cutthroat trout (GTNP)		$5,000			
Jackson Hole weed management education (GTNP)		$2,500			
Noxious weed Information and Education, Jackson Hole Wildlife Management Area (GTNP and BTNF)			$2,500		
Student Conservation Association Position— noxious weed program (GTNP)			$5,000		
Gros Ventre River spotted knapweed treatment (GTNP/ NER)			$5,000		
Protecting critical pronghorn migration corridor in southern GYA (GTNP)				$5,000	
Montana Conservation Corps, weed control and restoration work (GTNP)				$6,000	
Jackson Lake fishery trends (GTNP)				$5,000	
Sprinkler irrigation system in Taggart horse pasture (GTNP)				$5,000	
Grizzly, black bear–human interactions in the park (GTNP)					$7,500
Pronghorn population estimate summering in Jackson Hole area (GTNP)					$2,500
Visitor use assessment in Webb, Owl, and Berry Canyons (GTNP)					$5,000
Shoshone invasive species project (SNF)	$15,000	$15,000			

Project Title	Funded June 1, 2000	Funded March 26, 2001	Funded April 15, 2002	Funded March 5, 2003	Funded January 27, 2004
Whitebark pine planting (SNF)	$15,000				
Howard land exchange (SNF)		$5,000			
Winter use monitoring (SNF)		$6,000		$6,000	$6,000
Inland West Watershed Initiative watershed strategy (SNF)		$1,000			
Weed control, Wildlife Management Area support and initiation (SNF)			$9,000		
Whitebark pine blister rust plus tree selection program (SNF)			$3,500	$5,000	
Implementation of forestwide food storage/sanitation order (SNF)			$10,000	$7,500	
Grizzly bear food sources in the Wind River Range (SNF)				$3,000	
Togwotee Highway linkage zone analysis (SNF/BTNF)				$12,500	
Shoshone invasive species program (SNF)				$7,500	$7,500
Togwotee Highway movement study; food storage implementation (SNF)					$10,000
Yellowstone cutthroat trout mapping distribution (SNF)					$6,500
GYCC noxious weed inventory (YNP)	$5,000				
Distribution map for cutthroat trout and grayling (YNP)	$5,000	$10,000			
West Yellowstone visitor services (YNP through GNF)	$5,000				
Southwest backcountry weed control (YNP)		$7,500			
Whirling disease in cutthroat trout (YNP)		$15,000			
Lynx distribution surveys (YNP)		$5,000			

(*continued*)

Project Title	Funded June 1, 2000	Funded March 26, 2001	Funded April 15, 2002	Funded March 5, 2003	Funded January 27, 2004
Effects of wildfire on grizzly bears' vegetal foods (YNP)			$10,000		
Snowshoe hare inventory in YNP (YNP)			$7,500		
Army cutworm moth ecological studies (YNP)			$7,500		
Westslope cutthroat trout restoration in YNP (YNP)				$5,000	
GYA whitebark pine and blister rust surveys (YNP/GYA-wide)				$5,000	
Assessment of native cutthroat trout in the upper Yellowstone River–Thorofare region (YNP)				$7,500	
Whitebark pine plus tree program (YNP)					$5,000
Yellowstone/Snake River cutthroat inventory in the upper Snake River (YNP)					$7,500
Assessment of native cutthroat trout in the upper Yellowstone River (YNP)					$10,000
Elk, people, and plants: a century of change on the National Elk Refuge (NER)					$5,000
Hyper spectral remote sensing of whitebark pine (GYCC thru CNF)	$30,000	$10,000			
Executive director office support (GYCC through CTNF)	$5,000				
Interagency spatial analysis support (GYCC through GNF)	$20,000				
Ride the Right Trail (GYCC through GNF)	$10,000				
Yellowstone cutthroat trout viability analysis (GYCC through GNF)	$20,000				

Project Title	Funded June 1, 2000	Funded March 26, 2001	Funded April 15, 2002	Funded March 5, 2003	Funded January 27, 2004
Greater Yellowstone soil and landscape model (GYCC through GNF)	$20,000				
Acquisition of data services (GYCC)		$20,000	$7,000		
Weed awareness tools (GYCC)		$4,000			
GYA watershed vulnerability (GYCC)		$18,000			
Cutthroat trout documentary (GYCC)		$7,000			
Custer National Forest support costs (GYCC)		$7,500			
Wildland map and assessment (GYCC)			$9,000		
GYA-wide Forest Service Office Information and Education materials (GYCC)			$10,000	$7,500 (through SNF)	
Whitebark pine annotated bibliography (GYCC)			$1,000		
GYA whitebark pine blister rust studies (GYCC)			$10,000		
GYA noxious weed risk map (GYCC)			$8,000		
GYA reference stream reach (GYCC)			$30,000		
GYA digital fire history (GYCC)			$5,000		
GYA winter use monitoring (GYCC)			$30,000 (committed in 2001)		
GYA sand and gravel pit weed prevention inventory (GYCC)				$6,000	
GYA weed mapping (GYCC)				$18,000	
Vegetation database and GIS used for fire and fuels management (GYCC)				$10,000	
Recreation assessment (GYA)					$68,000
FARSITE data project (fire growth simulation model) (GYA)					$20,000

(continued)

Project Title	Funded June 1, 2000	Funded March 26, 2001	Funded April 15, 2002	Funded March 5, 2003	Funded January 27, 2004
Weed map (GYA)					$5,000
Project summary report (GYCC)					$10,000
GIS support (GYCC)					$2,000
Totals	48 submitted, 31 funded $310,000	56 submitted, 44 funded $300,000	64 submitted, 36 funded $300,000	81 submitted, 47 funded $285,000	69 submitted, 36 funded $288,500

Key to abbreviations: BDNF=Beaverhead-Deer Lodge National Forest, BTNF=Bridger-Teton National Forest, CNF=Custer National Forest, CTNF=Caribou-Targhee National Forest, GIS=Geographic Information System, GNF=Gallatin National Forest, GTNP=Grand Teton National Park, GYA=Greater Yellowstone Area, GYCC=Greater Yellowstone Coordinating Committee, NER=National Elk Refuge, SNF=Shoshone National Forest, YNP=Yellowstone National Park.

Appendix 4
Projects Funded by the Greater Yellowstone Coordinating Committee in 2001

Noxious Weed Management

Coordinated Noxious Weed Control Efforts

Unit	Project	Description	Partnerships
Beaverhead-Deerlodge National Forest	Madison Ranger District Backcountry Weed Management Program	Backcountry weed inventory, control, prevention, and education, primarily in the Gravelly Mountains	Rocky Mountain Elk Foundation; Madison Valley Ranchland Group; Madison County Weed Board
Bridger-Teton National Forest	Integrated Noxious Weed Control Project	Working with partners to control 150 acres of noxious weeds	Habitat Trust Fund; Wyoming Game and Fish Dept.; Rocky Mountain Elk Foundation; Jackson Hole Weed Management Area
Bridger-Teton National Forest	Purge Spurge	Eradicate leafy spurge along riparian areas in the Salt River Range of the Star Valley front	Highlands Cooperative Weed Management Lincoln County Area;
Custer National Forest	Rock Creek/Beartooth Highway Noxious Weed Project	Control spotted knapweed and leafy spurge that is expanding beyond the	Montana Dept. of Fish, Wildlife and Parks; Rock Creek Resort; City of Red Lodge;

(continued)

241

Unit	Project	Description	Partnerships
		Beartooth High-way right-of-way	Carbon County Weed District; 400 Ranch; Rock Creek Ranch
Caribou-Targhee National Forest	Black Canyon/Dry Canyon Leafy Spurge Control Program	Control spurge with sheep grazing and flea beetles on 2,500-acre infestation of spurge; monitor effectiveness	Utah/Idaho Weed Management Area; Jeffe Roche L&L Co.; Dry Canyon Cattlemen's Association; Sterling Bingham
Grand Teton National Park, National Elk Refuge, Bridger-Teton National Forest	Gros Venture River Corridor Spotted Knapweed Project	Continue project initiated last year to control knapweed seed sources located on major elk migration routes; expand control efforts to less accessible places	Teton Weed and Pest; Jackson Hole Weed Management Area

Cooperative Weed Management Areas (WMA), Education and Awareness

Unit	Project	Description	Partnerships
Gallatin National Forest	Upper Gallatin Invasive Species Project	Formally establish Upper Gallatin WMA; expand education, awareness and control efforts	Gallatin County Weed District; Montana Dept. of Fish, Wildlife, and Parks; Big Sky Institute
Gallatin National Forest	Upper Yellowstone Weed Management Program	Establish a WMA for the Upper Yellow-stone watershed	Park County; Bureau of Land Management; Montana Dept. of Fish, Wildlife, and Parks; Yellowstone National Park
Grand Teton National Park	Student Conservation Association (SCA) Position for Noxious Weed Education Campaign	Fund one student conservation associate for six months to work with Jackson Hole WMA on weed awareness	Teton County Weed and Pest; Student Conservation Association

Unit	Project	Description	Partnerships
GYCC all units	GYCC Weed Awareness Tools	Prepare internal awareness and training tools including Power Point presentations and Greater Yellowstone Area Weed Pocket Guide	Center for Invasive Plant Mgmt,; Montana State Univ.; Montana Weed Control Education Committee; Cooperative Ecosystems Study Unit

Inventory and Mapping Projects

Unit	Project	Description	Partnerships
Beaverhead-Deerlodge National Forest	Madison Valley Weed Inventory Project	Develop a comprehensive weed inventory for all lands; expand Upper Madison WMA to include entire valley	Madison County Ranchland Group; Madison County Weed Board; Bureau of Land Management
Caribou-Targhee National Forest	Palisades Wilderness Study Area Monitoring and Noxious Weed Control Project	Control noxious weed infestations discovered last year; inventory trails to relocate out of riparian areas; monitor recreation and grazing use	Upper Snake Weed Management Area; Bureau of Reclamation; Back Country Horsemen; Palisades Mitigation Project; Palisades Creek Ranch
Gallatin National Forest	Gardner Basin Noxious Weed Inventory	Map distribution and density of weeds, model potential spread on a key portion of the northern winter range	Foundation for North American Wild Sheep; Montana State University; Montana Dept. of Fish, Wildlife, and Parks
Shoshone National Forest	Shoshone Invasive Species Project	Complete inventory of forest, treat wilderness infestations, and establish Upper Wind River Weed Management Area	Bureau of Land Management; Fremont County; Cody Conservation District; South Central Wyoming College

(continued)

Unit	Project	Description	Partnerships
Yellowstone National Park	Southwest Yellowstone Backcountry Weed Survey and Control	Survey 10,000 acres for leafy spurge and other priority species; increase prevention, early detection, and containment along access roads	Idaho Dept. of Agriculture; Fremont County, Idaho; Henry's Fork Weed Management Area Caribou-Targhee National Forest

Wildlife Studies and Projects

Unit	Project	Description	Partnerships
Caribou-Targhee National Forest	Trumpeter Swan Nest Habitat Restoration Project	Improve water levels and nesting habitat for trumpeter swans nesting at four lakes (Swan, Beaver, Ernst, and Mesa Marsh)	Trumpeter Swan Society; Idaho Dept. of Fish and Game; Ducks Unlimited; U.S. Fish and Wildlife Service
Caribou-Targhee National Forest	Wolverine Natal Denning Habitat Mapping and Field Surveys	Run GIS model to map potential wolverine den habitat on Caribou National Forest; conduct additional aerial and ground surveys	Idaho Dept. of Fish and Game; Univ. of California at Santa Cruz
Gallatin National Forest	Effects of back country human use on bears and other large carnivores	Final year of a four-year project addressing the relationship of bears to human back country activities during hunting season	Interagency Grizzly Bear Study Team; Montana Dept. of Fish, Wildlife, and Parks; Hell's A-Roarin' Outfitters; student volunteers
Grand Teton National Park, Caribou-Targhee National Forest	Wolverine survey and monitoring	Evaluate wolverine habitat use and den selection in relation to human recreation use	Wildlife Conservation Society; Hornocker Institute; Wyoming Game and Fish Dept.; Idaho Dept. of Fish and Game; Alta 4-H Club; Grand Targhee Resort

Unit	Project	Description	Partnerships
Yellowstone National Park	Presence and distribution of lynx in Yellowstone National Park	Conduct intensive surveys in prime habitat using snow tracking and hair snares for DNA sampling	Yellowstone Park Foundation

Soil and Watershed Management

Unit	Project	Description	Partnerships
GYCC	Greater Yellowstone Soil and Landscape Model	Develop digital seamless map for landscape data (vegetation, soils, landforms, geology) for entire GYA at landscape scale	Partnership with units
GYCC	GYA Inland West Watershed Initiative Report	Report compiles GYA watershed data, identifies management strategies including restoration opportunities	Partnership with units
GYCC	GYA Inland West Watershed Vulnerability Report	Compile a consistent, reliable GIS or data layer that portrays vulnerability to disturbance	Partnership with units

Yellowstone Cutthroat Trout Conservation Efforts

Unit	Project	Description	Partnerships
GYCC all units	Prepare a Yellowstone cutthroat trout population viability assessment	Coordinated approach to complete a consistent GYA-wide viability assessment that summarizes current status and condition of populations	States of Idaho, Montana, Wyoming, all units
Custer National Forest, Gallatin National Forest	Yellowstone River Yellowstone Cutthroat Distribution Study	Survey Yellowstone River tributaries on Custer and Gallatin National Forests to collect information on distribution and genetic status of trout populations	Montana Dept. of Fish, Wildlife, and Parks; Montana State University

(*continued*)

Unit	Project	Description	Partnerships
Gallatin National Forest	Gallatin River Basin westslope cutthroat trout reintroduction	Collection and analysis of base-line data for currently known or suspected fishless stream reaches in the Gallatin River basin	Montana Dept. of Fish, Wildlife, and Parks; Bozeman Watershed Council; Yellow-stone National Park
Grand Teton National Park	Effects of irrigation ditches on water quality/cutthroat trout habitat on Snake River tributaries	Determine effects of irrigation on water quality and cutthroat trout habitat;assess need for mitigation	Wyoming Game and Fish Dept.
Caribou-Targhee National Forest	Role of beavers in trout habitat and hydrologic function	Inventory Teton River drainage to see if beavers can help restore watershed function and health	Idaho Div. of Environmental Quality; Idaho Fish and Game; Teton Soil Conservation District; Natural Resource Conser-vation Service
Caribou-Targhee National Forest	Yellowstone cutthroat trout distribution/ habitat surveys and mapping	Complete an additional 30 surveys in the Snake River drainage to deter-mine Yellowstone cutthroat distribution and habitat quality; 20 new populations were inventoried in 2000	U.S. Bureau of Reclamation; Federation of Fly Fishers; Idaho Dept. of Fish and Game; University of Idaho; Trout Unlimited; U.S. Fish and Wildlife Service
Yellowstone National Park	Yellowstone National Park distribution map for cutthroat trout and grayling	Develop historic and current GIS distribution layers for Yellowstone and westslope cutthroat trout and grayling	Coordinating with Shoshone, Gallatin, and Bridger-Teton National Forests
Yellowstone National Park	Determination of distribution and severity of whirling disease in Yellow-stone cutthroat	Time series exposure tests of vulnerable cutthroat fry conducted at eight sites; examine	Montana Dept. of Fish, Wildlife, and Parks; Wyoming Game and Fish Dept.;

Unit	Project	Description	Partnerships
	trout in Yellowstone Lake Basin	older cutthroat for signs of the disease	U.S. Fish and Wildlife Service Bozeman Fish Health Lab
Caribou-Targhee National Forest	Pine Creek fish weir, South Fork of the Snake River	Install weir to prevent nonnative fish from traveling upstream to spawn with native cut-throat trout	Idaho Dept. of Fish and Game; Trout Unlimited; One Fly Foundation

Education and Awareness

Unit	Project	Description	Partnerships
Custer National Forest, Gallatin National Forest	Yellowstone cutthroat trout interpretive display	Develop mobile interpretive dis-play portraying current status and conservation efforts for Yellow-stone cutthroat trout	
GYCC all units	Yellowstone cutthroat trout interpretive video/documentary	Continue with FY 2000 project to develop a cutthroat trout documentary film for viewing on TV and for inter-pretive efforts	Potential partners include nine states, five federal agencies, three foundations, and three conserva-tion groups

Whitebark Pine Restoration and Management

Unit	Project	Description	Partnerships
Caribou-Targhee National Forest	Whitebark pine planting	Plant 20 acres of whitebark pine	Global Releaf; National Arbor Day Foundation
Gallatin National Forest	Whitebark pine planting	Plant 10 acres of whitebark pine on lands acquired from Big Sky Lumber	National Arbor Day Foundation
GYCC all units	Hyperspectral remote sensing of whitebark pine	Analyze data collected in FY 2000 and prepare final report addressing feasibility for future inventory and monitoring efforts	Yellowstone Ecosystem Studies; U.S. Geological Survey; Biological Resources Division, Intera-gency Grizzly Bear Study Team

(*continued*)

Unit	Project	Description	Partnerships
Shoshone National Forest	Whitebark pine planting	Plant 20 acres of whitebark pine	National Arbor Day Foundation; Plant a Tree Foundation

Land Patterns

Unit	Project	Description	Partnerships
Beaverhead-Deerlodge National Forest	Gravelly Mountains land exchange	Fund staff work to complete land exchange to acquire 219 acres of private land in the Gravelly Mountains	
Gallatin National Forest	Royal Teton Ranch (RTR) lands work	Fund staff work associated with easements, rights-of-way, permits, and boundary management	Rocky Mountain Elk Foundation
Gallatin National Forest	Duck Creek wetlands—critical land acquisition	Forest is pursuing purchase of property; funding will provide for continued negotia-tions to develop a purchase option	
Gallatin National Forest	Slip and Slide Ranch conservation easement	Negotiate and secure a conservation easement for 700 acres of key winter range	Rocky Mountain Elk Foundation; Montana Dept. of Fish, Wildlife, and Parks
Gallatin National Forest	Historic OTO Ranch Management Plan	Develop a manage-ment plan for historic ranch compatible with northern winter range and threatened and endangered habitat	Amizade Ltd. and Elder Hostel helping with restoration
Shoshone National Forest	Howard Land exchange, South Fork of the Shoshone River	Help fund necessary staff work to complete land exchange to improve land patterns in the South Fork Shoshone River	Nature Conservancy

Unit	Project	Description	Partnerships
Recreation and Visitor Services			
All national forests	Ride the Right Trail	Develop strategy, educational material, and signing to encourage ethical use of off-highway vehicles	
Bridger-Teton National Forest	Teton Division visitor support	Increase summer staffing and presence at Blackrock Ranger Station, a major portal into Grand Teton and Jackson Hole	
Bridger-Teton National Forest	Teton Wilderness Salt Site Study, Phase 2	Sample an additional 30 salting sites, conduct analysis, produce final and reclamation report	University of Montana; other groups will help with restoration work in the future
Bridger-Teton National Forest	Resort naturalist program	Partnership with 10 local resorts and dude ranches to provide interpretive programs	10 resorts; Snake River Fund; 2 Eagle Scout projects
Custer National Forest, Gallatin National Forest, Shoshone National Forest	Beartooth Scenic Byway Corridor Management Plan	A corridor management plan will coordinate interpretive and recreation opportunities and is needed to compete for All American Road designation	Yellowstone Country; Rocky Fork Ranch; Red Lodge Lodging Association; Red Lodge Chamber of Commerce; Cody Chamber of Commerce; Park County Travel Council
Gallatin National Forest, Yellowstone National Park	West Yellowstone public lands information desk	Provides funding necessary to staff key entry portal in to the Greater Yellowstone Area at West Yellowstone	Town of West Yellowstone

(*continued*)

Unit	Project	Description	Partnerships
All national forests	Winter use monitoring	Implement a coordinated monitoring program for 6 national forests to collect information on use trends and where use is occurring	Cooperating with states of Montana, Idaho, and Wyoming on data collection

Source: Greater Yellowstone Coordinating Committee, "GYCC at Work," *Greater Yellowstone Coordinating Committee Briefing Guide* (Billings, Mont.: Greater Yellowstone Coordinating Committee, 2001), 10–13.

Appendix 5
The Tasks of Problem Solving

There are five required, interactive problem-solving tasks. The first task is clarification of goals (values), sometimes called the normative standpoint. It must end with a detailed value statement of goals and objectives and should make clear the empirical reference to the preferred events in the social process. Indices should be established and tracked quantitatively and qualitatively. Goal clarification should build on knowledge of past data about trends in decision making, conditions under which problems arose, and likely future probabilities. In greater Yellowstone the goal at an abstract level might be to manage the region's natural and cultural resources sustainably in ways that enjoy broad public support. In turn, a lower order and more specific objective might be to develop highly effective, efficient, and equitable interagency coordination and integration as one means to achieve the higher order goal. Goal clarification demands serious effort. This task needs to be revisited repeatedly through the life cycle of any management policy, whether it is for weed management in greater Yellowstone or climate change worldwide.

The second task is description of past trends, sometimes called the *historical standpoint*. This standpoint must be taken by problem solvers to determine the past trends or history in the problem at hand, in social interactions, and in decision making. These historic trends should be described in terms of whether they approximate, in terms of constitutive process and common interest, the policies and goals already clarified in the policy (value) statement. These historical matters must be detailed factually in terms of indices that are tangible and measurable, and they must be described in terms of the human value processes that produced them.

The third task is analysis of the conditioning factors, or coming up with an explanation of why the problem is a problem, why social interaction is problematic, and how decision making has unfolded as it has. This task is sometimes called the *scientific* or *explanation standpoint*. Conceptions and theories to "explain" observed outcomes should be advanced and tested by appropriate methods of modern science, both social and biophysical sciences. Theories should be based on the "maximization postulate," which says that people behave in ways that they perceive will leave them better off in terms of value "indulgences" than the alternatives. People's perceptions are based on predispositional (personality) and environmental (external) variables, and people can and do "misperceive." An individual's behavior is a product of many factors; culture, class, interest, personality, and crises should be explicitly examined as possible causes of decisions. Leaders can try to be rigorous in analyzing the conditions of important trends, but they must take care not to overemphasize theoretical models, mathematical measurement, or experimentation to explain history.

The fourth task is projection, sometimes called the *futuring* or *forecasting standpoint*. This task looks at likely future trends in terms of the biophysical problem at hand, in terms of human interactions, and in decision-making processes and effects. People's expectations about the future should be made as clear, conscious, and realistic as possible. Simple linear or chronological extrapolations using conventional theories are grossly inadequate for this demanding and creative task. Leaders can try to build "developmental constructs"—educated projections based on knowledge of goals, trends, and conditions, using alternative notions of the future. Scenarios are often used to fulfill this task. At last, three "futures" should be outlined (best case, expected case, and worst case), and these should be tested in light of all available information. Projections should be subject to disciplined knowledge about historical changes and conditioning factors. Completing this task lets the problem solver know if a discrepancy exists between goals and likely future events. If there is a discrepancy, it means that a problem exists and so a problem definition must be created. Completing this task makes it possible for leaders to determine what the contours of the problem are and how it might be addressed.

The fifth and final task is the invention, evaluation, and selection of alternatives to solve the problem as defined. This is sometimes called the *practical* or *operational standpoint*. Creativity is a crucial element in inventing and evaluating alternative solutions. Leaders must determine which deliberate interventions will be most effective in solving the problem at hand. Every phase of the decision process and every aspect of the conditioning context should be examined for opportunities to influence decision making. There are only five variables that can be changed for better outcomes—participants, their perspectives, the situation, the values at play, and participants' strategies. The proposed solutions must target one or more of these. Leaders must also make realistic assessments of alternatives in terms of gains or losses with respect to all eight values. Knowing how values might change among participants under each alternative being considered is a vital part of making a final selection for implementation. Leaders must make a final determination based on disciplined knowledge of goals, trends, conditions, and future possibilities.

Notes

Chapter 1. Leaders and Policy in a Contested Landscape

1. C. Carroll as cited in W. Stolzenburg, "The Long Rangers," *Nature Conservancy* 53, no. 3 (2003): 38.

2. T. Kerasote, *Heart of Home: People, Wildlife, and Place* (Guilford, Conn.: Lyons Press, 2003).

3. This case is covered in Editor, "Forest Should Reconsider Leases," *Jackson Hole News and Guide,* September 1, 2004, A2; R. Hunting, "Governor to Tour Wyoming Range Leases," *Jackson Hole News and Guide,* September 1, 2004, A22; R. Huntington, "Forest Has Authority to Stop Gas Leasing," *Jackson Hole News and Guide,* September 1, 2004, A23; Editor, "Gov's Plea Worthy," *Jackson Hole News and Guide,* September 15, 2004, A4; J. Stanford, "Forest Halts Energy Leases," *Jackson Hole News and Guide,* September 15, 2004, A1; T. Wilkinson, "Tell Us Kniffy: Who Wants You to Drill?" *Jackson Hole News and Guide,* September 22, 2004, A6; T. Wilkinson, "Forest Is Backward with Energy Leasing," *Jackson Hole News and Guide,* September 29, 2004, A6; L. Dorsey, "In Defense of Kniffy," *Jackson Hole News and Guide,* September 29, 2004, A30; and political cartoon, *Jackson Hole News and Guide,* October 6, 2004, A3.

4. There is much controversy on the proposed bike path part of the transportation plan. Newspapers are full of articles pro and con, such as R. Huntington, "Group Plugs Report by Pathways Expert," *Jackson Hole News and Guide,* August 10, 2005, A10. Many organizations in the region oppose it.

5. See U.S. Fish and Wildlife Service and National Park Service, "Draft Bison and Elk Management Plan and Environmental Impact Statement," July 20, 2005, Denver, Colo.

6. Editor, "Feds Fail Public," *Jackson Hole News and Guide,* August 10, 2005, A4.

7. P. Schullery, *Searching for Yellowstone: Ecology and Wonder in the Last Wilderness* (Boston: Houghton Mifflin, 1997).

8. R. K. Yin, 1994, *Case Study Research: Design and Methods* (Thousand Oaks, Calif.: Sage), 13; and T. W. Clark, "Case Studies in Wildlife Policy Education," *Renewable Resources Journal* (Autumn 1986): 11–17.

9. T. W. Clark, *The Policy Process: A Practical Guide for Natural Resource Professionals* (New Haven: Yale University Press, 2002), provides an overview of this approach and framework. See also R. D. Brunner, "Introduction to the Policy Sciences," *Policy Sciences* 30, no. 4 (1997): 191–215; R. D. Brunner, "Teaching the Policy Sciences: Reflection on a Graduate Seminar," *Policy Sciences* 30, no. 4 (1997): 217–31.

10. E. Babbie, 2001, *The Practice of Social Research* (Belmont, Calif.: Wadsworth/Thompson), 113.

11. T. W. Clark and S. C. Minta, *Greater Yellowstone's Future: Prospects for Ecosystem Science, Management, and Policy* (Moose, Wyo.: Homestead Press, 1994).

12. See M. H. Arsanjani, *International Regulation of Internal Resources: A Study of Law and Policy* (Charlottesville: University of Virginia Press, 1981). This section is based on Arsanjani's look at natural resource management policy.

13. The concept of "interests" is a complex one. Perhaps the best introduction to this concept and how interests play out in real life is by M. S. McDougal and W. M. Reisman, eds., *International Law Essays: A Supplement to International Law in Contemporary Practice* (Mineola, N.Y.: Foundation Press, 1981).

14. A case is documented in G. Ferguson, *Hawk's Rest: A Season in the Remote Heart of Yellowstone* (Washington, D.C.: National Geographic Adventure Press, 2003).

15. National Research Council, *Our Common Journey: A Transition Toward Sustainability* (Washington, D.C.: National Academy Press, 1999), 1.

16. For a discussion of sustainability, see N. Mirovitskaya and W. Ascher, *Guide to Sustainable Development and Environmental Policy* (Durham: Duke University Press, 2001).

17. M. Flader, U. M. Goodale, A. G. Lanfer, C. Margoulis, and M. Stern, eds., *Transboundary Protected Areas: The Viability of Regional Conservation Strategies. Journal of Sustainable Forestry* 17, nos. ½ (2003). This is a special issue on transboundary issues.

18. D. Zbicz, "Imposing Transboundary Cooperation between Internationally Adjoining Protected Areas," in M. Flader et al., eds., *Transboundary Protected Areas: The Viability of Regional Conservation Strategies, Journal of Sustainable Forestry* 17, nos. ½ (2003): 21–38.

19. K. Sherman and M. Duda, "An Ecosystem Approach to Global Assessment and Management of Coastal Waters," *Marine Ecology Program Series* 190 (1999): 271–87.

20. Zbicz, "Imposing Transboundary Cooperation."

21. M. S. McDougal, "Legal Basis for Securing the Integrity of the Earth-Space Environment," *Journal of Natural Resources and Environmental Law* 8, no. 2 (1992–93): 177–207. See also M. S. McDougal, H. D. Lasswell, and L. Chen, *Human Rights and World Public Order: The Basic Polices of an International Law of Human Dignity* (New Haven: Yale University Press, 1980).

22. See, for example, H. D. Lasswell, "From Fragmentation to Configuration," *Policy Sciences* 2 (1971): 439–46; L. H. Gunderson, C. S. Holling, and S. L. Light, "Barriers Broken and Bridges Built: A Synthesis," in L. H. Gunderson, C. S. Holling, and S. L. Light, eds., *Barriers and Bridges to the Renewal of Ecosystems and Institutions* (New York: Columbia University Press, 1995), 489–532; National Research Council, "Integrating Knowledge and Action," *Our Common Journey: A Transition Toward Sustainability* (Washington, D.C.: National Academy of Sciences Press, 1999), 275–332; and C. Folke, F. Berkes, and J. Colding, "Ecological Practices and Social Mechanisms for Building Resilience and Sustainability," in C. Folke, F. Berkes, and J. Colding, eds., *Linking Social and Ecological Systems: Management*

Practices and Social Mechanisms for Building Resilience (New York: Cambridge University Press, 2000), 414–36. See also G. D. Brewer, ed., "The Theory and Practice of Interdisciplinary Work," *Policy Sciences* 32, no. 4 (1999): 315–429.

23. For example, see R. L. Knight, and P. B. Landres, eds., *Stewardship Across Boundaries* (Washington, D.C.: Island Press, 1998). In this volume, specifically see D. A. Glick and T. W. Clark, "Overcoming Boundaries: The Greater Yellowstone Ecosystem," 237–56. Also see C. W. Thomas, *Bureaucratic Landscapes: Interagency Cooperation and the Preservation of Biodiversity* (Cambridge: MIT Press, 2003); Sherman and Duda, "Ecosystem Approach to Global Assessment"; and transcons-l@listproc.tufts.edu.

24. For "thinking outside the box," C. A. Schön, *The Reflective Practitioner: How Professionals Think in Action* (New York: Basin Books, 1983), is a good introduction. The antidote to "boxed in" thinking is outlined in T. W. Clark, M. B. Rutherford, K. Ziegelmayer, and M. J. Stevenson, "Conclusion: Knowledge and Skills for Professional Practice," in T. W. Clark, M. Stevenson, K. Ziegelmayer, and M. B. Rutherford, eds., *Species and Ecosystem Conservation: An Interdisciplinary Approach*, Bulletin No. 105 (New Haven: Yale School of Forestry and Environmental Studies, 2001), 253–376.

25. One good example of this strategy is S. Primm and S. M. Wilson, "Reconnecting Grizzly Bear Populations: Prospects for Participatory Projects," *Ursus* 15, no. 1 (2004): 104–14.

26. For examples of large-scale conservation efforts in other environments, see S. B. Olsen and D. Nickerson, *The Governance of Coastal Ecosystems at the Regional Scale: An Analysis of the Strategies and Outcomes of Long-term Programs*, Coastal Management Report No. 2243 (Narragansett: University of Rhode Island Coastal Resources Center, 2003); Sherman and Duda, "Ecosystem Approach to Global Assessment"; M. Schultz and E. Yaghmour, eds., *Waiting for Democracy: The Politics of Choice in Natural Resource Decentralization* (Washington, D.C.: World Resources Institute, 2004).

27. T. R. Stanley, "Ecosystem Management and the Arrogance of Humanism," *Conservation Biology* 9, no. 2 (1995): 256 (all quotes).

28. R. E. Grumbine, "What Is Ecosystem Management?" *Conservation Biology* 8, no. 1 (1994): 31.

29. R. D. Brunner and T. W. Clark, "A Practice-Based Approach to Ecosystem Management," *Conservation Biology* 11, no. 1 (1997): 48–58.

30. M. E. Jensen and P. S. Bourgeron, eds., *Ecosystem Management: Principles and Applications,* vol. 2 (U.S. Department of Agriculture, Forest Service, Pacific Northwest Research Station, General Technical Report PNW-GTR-318, 1994), 3.

31. Brunner and Clark, "Practice-Based Approach to Ecosystem Management."

32. R. L. Caldwell, "The Ecosystem as a Criterion for Public Land Policy," in R. L. Smith, ed., *The Ecology of Man: An Ecosystem Approach* (New York: Harper and Row, 1972), 413.

33. H. M. Rauscher, "Ecosystem Management Decision Support for Federal Forests in the United States: A Review," *Forest Ecology and Management* 114 (1999): 173–97, 173.

34. Rauscher, "Ecosystem Management Decision Support," 173. See also G. Swogger, Jr., "The Psychodynamics of Threat in Environmental Rhetoric," *Technology* 6 (1999): 173–91.

35. M. B. Rutherford, "Conceptual Frames, Values of Nature, and Key Symbols of Ecosystem Management: Landscape Scale Assessment in the Bridger-Teton National Forest" (Ph.D. diss., Yale School of Forestry and Environmental

Studies, 2002). See also H. J. Cortner and M. A. Moote, *The Politics of Ecosystem Management* (Washington, D.C.: Island Press, 1999); and T. W. Clark, "Creating and Using Knowledge for Species and Ecosystem Conservation: Science, Organizations, and Policy," *Perspectives in Biology and Medicine* 36 (1993): 497–525, appendixes.

36. U.S. General Accounting Office, *Ecosystem Management: Additional Actions Needed to Adequately Test a Promising Approach,* USGAO Publication #GAO/RCED-94-111 (Washington, D.C.: USGAO, 1994).

37. S. L. Yaffee, A. F. Phillips, I. C. Frentz, P. W. Hardy, S. M. Maleki, and B. E. Thorpe, *Ecosystem Management in the United States: An Assessment of Current Experiences* (Washington, D.C.: Island Press, 1994), 4.

38. R. Haeuber, "Setting the Environmental Policy Agenda: The Case of Ecosystem Management," *Natural Resources Journal* 36 (1996): 1–27.

39. These are introduced in H. D. Lasswell and A. Kaplan, *A Framework for Political Inquiry* (New Haven: Yale University Press, 1950) and in the many other works by Lasswell and his colleagues cited in this volume. There are many other partial notions of these concepts and tools in use presently; none is as comprehensive or practical as those put forth by Lasswell. In the worst situations, leaders lose their integrity and personal responsibility. They become more interested in furthering their individual careers than anything else. As a consequence, organizations lose their effectiveness. One example where this happened is described in R. A. Gabriel and P. L. Savage, *Crisis in Command: Mismanagement in the Army* (New York: Hill and Wang, 1978).

40. This skill set originally comes from H. D. Lasswell and M. S. McDougal, *Jurisprudence for a Free Society: Studies in Law, Science, and Policy* (New Haven: New Haven Press, 1992). It was adapted by T. W. Clark, "Developing Policy-Oriented Curricula for Conservation Biology: Professional and Leadership Education in the Public Interest," *Conservation Biology* 15, no. 1 (2000): 31–39; and T. W. Clark and M. Ashton, "Interdisciplinary Rapid Field Appraisals: The Ecuadorian Condor Bioreserve Experience," *Journal of Sustainable Forestry* 18, no. 2/3 (2004): 1–25. To read further on leadership skills, see Clark et al., "Conclusion: Knowledge and Skills for Professional Practice"; and T. W. Clark, A. R. Willard, and C. M. Cromley, eds., *Foundations of Natural Resources Policy and Management* (New Haven: Yale University Press, 2000); and Clark, *Policy Process*.

41. W. Ascher and B. Hirschfelder-Ascher, "Political Psychology and the Risks of Leadership," in W. Ascher and B. Hirschfelder-Ascher, *Revitalizing Political Psychology: The Legacy of Harold D. Lasswell* (Mahwah, N.J.: Lawrence Erlbaum Associates, 2005), 95–115.

42. J. M. Burns, *Leadership* (New York: Harper Torchbooks, 1978).

43. Other scholars and practitioners of leadership would agree, but offer their own views on the subject, including J. W. Gardner, *On Leadership* (New York: Free Press, 1990); V. H. Vroom and A. G. Jago, *The New Leadership: Managing Participation in Organizations* (Englewood Cliffs, N.J.: Prentice-Hall, 1988); J. J. Cribbin, *Leadership: Strategies for Organizational Effectiveness* (New York: American Management Association, 1981); and G. Wills, "What Makes a Good Leader?" *Atlantic Monthly,* April 1994, 63–80.

44. J. R. McNeill, *Something New Under the Sun: An Environmental History of the Twentieth-Century World* (New York: W. W. Norton, 2000).

45. P. F. Drucker, "The Age of Social Transformation," *Atlantic Monthly,* November 1994, 53.

46. L. C. Thurow, "Building Wealth," *Atlantic Monthly,* June 1999, 57.

47. J. A. Beesley and M. S. McDougal, "Foreword," in J. Schneider, *World Public Order of the Environment: Towards an International Ecological Law and Organization* (Toronto: University of Toronto Press, 1979), viii.

48. Worldwatch Institute, *Vital Signs* (New York: W. W. Norton, 2002), 88.

49. Ibid., 59.

50. J. G. Speth, *Red Sky at Morning: America and the Crises of the Global Environment* (New Haven: Yale University Press, 2004), 20–21.

51. Congress on Promoting Sustainability in the Twenty-first Century, "Tools and Strategies for Sustainability," special report, *Renewable Resources Journal* 19, no. 1 (2001): 16, quote on p. 16.

52. F. H. Bormann, *Confronting the Environmental Debt* (New Haven: Yale University Press, 1997).

53. V. H. Heywood, ed., *Global Biodiversity Assessment*, United Nations Environmental Program (New York: Cambridge University Press, 1995).

54. World Resources Institute et al., *World Resources, 2000–2001* (Washington, D.C.: World Resources Institute, 2001), 3–4.

55. H. D. Lasswell, "Future of Government and Politics in the United States," in L. Rubin, ed., *The Future of Education: Perspectives on Tomorrow's Schooling* (Boston: Allyn and Bacon, 1975), 1–21.

56. N. Ferguson, "2011," *New York Times Magazine*, December 2, 2001, 76.

57. M. Brzezinski, "Fortress America," *New York Times Magazine*, February 23, 2003, 38.

58. M. J. Sandel, "America's Search for a New Public Philosophy," *Atlantic Monthly*, March 1996, 57.

59. B. Schwarz and C. Layne, "A Grand New Strategy," *Atlantic Monthly*, January 2002, 36.

60. E. O. Wilson, *The Diversity of Life* (Cambridge: Belknap Press, 1992), 20–21; and U.S. Geological Survey, *Status and Trends of the Nation's Biological Resources* (Washington, D.C.: U.S. Government Printing Office, 1998); and H. John Heinz III Center for Science, Economics, and the Environment, *The State of the Nation's Ecosystems: Measuring the Lands, Waters, and Living Resources of the United States* (New York: Cambridge University Press, 2002).

61. See Sonoran Institute Web page, www.sonoran.org. For a broad view of regional culture, see S. Western, *Pushed Off the Mountain, Sold Down the River: Wyoming's Search for Its Soul* (Moose, Wyo.: Homestead Publishing, 2002); M. P. Malone, *Montana: A Contemporary Profile* (Helena: American and World Geographic Publishing, 1996); and W. Stegner, *The American West as Living Space* (Ann Arbor.: University of Michigan Press, 1987).

62. R. D. Brunner, "Myth and American Politics," *Policy Sciences* 27 (1994): 1–18.

63. See the League of Conservation Voters, www.lcv.org (accessed November 22, 2004).

64. See P. Lichtman and T. W. Clark, "Rethinking the 'Vision' Exercises in the Greater Yellowstone Ecosystem," *Society and Natural Resources* 7 (1994): 459–78; and S. A. Primm and T. W. Clark, "The Greater Yellowstone Policy Debate: What Is the Policy Problem?" *Policy Sciences* 29 (1996): 137–66.

65. R. D. Brunner, "The Policy Movement as a Policy Problem," *Policy Sciences* 24 (1991): 65–98.

66. D. Katz and R. L. Kahn, *The Social Psychology of Organizations* (New York: John Wiley and Sons, 1978).

67. For a partial listing of ordinary challenges, see Yellowstone National Park, *Yellowstone Resources Issues: An Annual Compendium of Information About*

Yellowstone National Park (Mammoth, Wyo.: Yellowstone National Park, 2001); Greater Yellowstone Coalition, *Status of Fisheries and Aquatic Habitats in the Greater Yellowstone Ecosystem* (Bozeman: Greater Yellowstone Coalition, 1999). See also A. M. Thuermer, Jr., "Yellowstone Declared 'in Danger,' " *Jackson Hole News,* December 6, 1995, A1; and D. Quammen, "National Parks: Nature's Dead End," *New York Times,* July 28, 1996, E13; finally, read World Resources Institute, *People and Ecosystems: The Fraying Web of Life* (Washington, D.C.: World Resources Institute, 2000–2001).

68. Cortner and Moote, *On the Politics of Ecosystem Management,* xi.

69. See C. M. Cromley, "Bison Management in Greater Yellowstone," and R. D. Brunner, "Problems of Governance," in R. D. Brunner, C. H. Colburn, C. M. Cromley, and R. A. Klein, eds., *Finding Common Ground: Governance and Natural Resources in the American West* (New Haven: Yale University Press, 2002), 126–58 and 1–47. This book contains additional Yellowstone cases. W. McKinney and W. Harmon, in *The Western Confluence: A Guide to Governing Natural Resources* (Washington, D.C.: Island Press, 2004), also see that basic problems are rooted in our governance structure.

70. These are some key papers on the constitutive process. Perhaps most comprehensive is M. S. McDougal, H. D. Lasswell, and W. M. Reisman, "The World Constitutive Process of Authoritative Decision," in M. S. McDougal and W. M. Reisman, eds., *International Law Essays: A Supplement of International Law in Contemporary Perspective* (Mineola, N.Y.: Foundation Press, 1981), 191–286. The following references are also important: Lasswell, "Professional Services: The Constitutive Policy Process," in *Pre-view of Policy Sciences* (New York: Elsevier, 1971), 98–111; McDougal, Lasswell, and Reisman, "Theories about International Law: Prologue to a Configurative Jurisprudence," in *International Law Essays,* 43–141 and see 56; and Reisman, "Law From a Policy Perspective," in *International Law Essays,* 1–14 (see "The Constitutive Process," 9–14).

71. Lasswell and McDougal, *Jurisprudence for a Free Society.*

72. R. G. Healy and W. Ascher, "Knowledge in the Policy Process: Incorporating New Environmental Knowledge in Natural Resource Policy Making," *Policy Sciences* 28 (1995): 1–19.

73. M. S. McDougal and W. T. Burke, "The Public Order of the Oceans: A Contemporary International Law of the Sea (New Haven: Yale University Press, 1962; repr. 1987, with new introductory essay); M. S. McDougal, H. D. Lasswell, and I. A. Vlasic, *Law and Public Order in Space* (New Haven: Yale University Press, 1963); D. M. Johnston, *The International Law of Fisheries: A Framework for Policy-Oriented Inquiries* (New Haven: Yale University Press, 1965; repr. 1987, with new introductory essay); E. J. Sahurie, *The International Law of Antarctica* (New Haven: Yale University Press, 1992). See also Arsanjani, *International Regulation of Internal Resources.*

74. A. Leopold, "The Land Ethic" in *A Sand County Almanac: With Essays from Round River* (New York: Ballantine Books, 1966), 237–64.

75. Leopold, "Land Ethic," 3.

76. Brunner, "Policy Movement as a Policy Problem"; and Brunner, "Myth and American Politics."

77. J. F. Franklin, "Old-Growth Forests, Owls, and Conservation Paradigms," *Society for Conservation Biology Newsletter* 11, no. 3 (2004): 1.

78. National Research Council, *Our Common Journey: A Transition toward Sustainability,* 10. See also E. A. Parson and W. C. Clark, "Sustainable Development as Social Learning: Theoretical Perspective and Practical Challenges for the Design of a Research Program," and D. N. Michael, "Barriers and Bridges to Learning in a Turbulent Human Ecology," both in Gunderson et al., eds., *Barriers and Bridges to*

the Renewal of Ecosystems and Institutions, 428–60, 461–88. A method for doing this is detailed in Clark, *Policy Process.*

Chapter 2. Challenges Facing Greater Yellowstone

1. H. D. Lasswell and M. S. McDougal, *Jurisprudence for a Free Society: Studies in Law, Science, and Policy* (New Haven: New Haven Press, 1992), 425.

2. See J. L. Pressman and A. Wildavsky, *Implementation: How Great Expectations in Washington Are Dashed in Oakland: Or, Why It's Amazing That Federal Programs Work at All* (Berkeley: University of California Press, 1972); R. T. Nakamura and F. Smallwood, *The Politics of Policy Implementation* (New York: St. Martin's Press, 1980); and D. A. Mazmanian and P. A. Sabatier, *Implementation and Public Policy* (Glenview, Ill.: Scott, Foresman, 1983).

3. This is how T. W. Clark and S. C. Minta, *Greater Yellowstone's Future: Prospects for Ecosystem Science, Management, and Policy* (Moose, Wyo.: Homestead Press, 1994), define greater Yellowstone. Most environmental groups (e.g., Greater Yellowstone Coalition) define the ecosystem in similar ways, as do most academics, and even some agency staff members. There is, however, quite a range in how people define the actual boundaries and how porous those boundaries are.

4. J. C. Scott, *Seeing Like a State: How Certain Schemes to Improve the Human Condition Have Failed* (New Haven: Yale University Press, 1998). Agency systems tend to ignore the fact that local, practical knowledge is as important as formal, epistemic knowledge. Agency behavior can harm the complex interdependencies that exist in a region and among people independent of government. Agencies tend to impose administrative order on nature and society, employ a high modernist ideology that believes that scientific intervention can improve every aspect of human life. They also show a willingness to use authoritarian state power to effect large-scale changes and engineer a weak civil society that cannot effectively resist such efforts. See a review of this book by A. R. Willard, "Scott, James C., *Seeing Like a State: How Certain Schemes to Improve the Human Condition Have Failed,*" book review, *Policy Sciences* 33, no. 1 (2000): 107–15.

5. The political culture is a major determinant of the structure of the arena and decision making. See D. J. Elazar, *The American Mosaic: The Impact of Space, Time, and Culture on American Politics* (Boulder: Westview Press, 1994); J. Garreau, *The Nine Nations of North America* (Boston: Houghton Mifflin, 1981); R. D. Gastil, *Cultural Regions of the United States* (Seattle: University of Washington Press, 1975); D. J. Elazar, *American Federalism* (New York: Thomas Y. Cromwell, 1972). Also see B. Hall and M. L. Kerr, *Green Index: A State by State Guide to the Nation's Environmental Health* (Washington, D.C.: Island Press, 1991–1992); and W. P. Pendley, *War on the West: Government Tyranny on America's Great Frontier* (Washington, D.C.: Regency, 1995).

6. S. K. Arora and H. D. Lasswell, *Political Communication: The Public Language of Political Elites in India and the United States* (New York: Holt, Rinehart and Winston, 1969).

7. J. W. Wilmot and T. W. Clark, "Wolf Restoration: A Battle in the War over the West," in T. W. Clark, M. B. Rutherford, and D. Casey, eds., *Coexisting with Large Carnivores: Lessons from Greater Yellowstone* (Washington, D.C.: Island Press, 2005), 138–74.

8. Much has been written on agency rivalries. One of the best accounts is J. N. Clarke's and D. McCool's classic 1985 book, *Staking Out the Terrain: Power Differentials among Natural Resource Management Agencies* (Albany: State University of New York Press, 1985). It offers an inside look at the power rivalries among gov-

ernment agencies. The authors note that "whether policies succeed or fail in their objectives is largely dependent upon the nature of the organization mandated to carry out those policies." They go on to discuss "differentials in agency power" and why these differentials are so important in the policy process (p. 2).

9. U.S. Congress, House Committee on Interior and Insular Affairs, Subcommittee on Public Lands, *Greater Yellowstone Ecosystem,* oversight hearing before the Subcommittee on Public Lands and the Subcommittee on National Parks and Recreation of the Committee on Interior and Insular Affairs, House of Representatives, 99th Cong. 1st sess., hearing held in Washington, D.C., October 24, 1985 (Washington, D.C.: U.S. Government Printing Office, 1986); and *Greater Yellowstone Ecosystem: An Analysis of Data Submitted by Federal and State Agencies* (Washington, D.C.: U.S. Government Printing Office, 1987).

10. U.S. Congress, House Committee on Interior and Insular Affairs, Subcommittee on Public Lands, *Greater Yellowstone Ecosystem,* oversight hearing, 4.

11. Many of these works, authors, and problems are summarized in T. W. Clark, A. H. Harvey, M. B. Rutherford, B. Suttle, and S. A. Primm, comp., *Management of the Greater Yellowstone Ecosystem: An Annotated Bibliography,* 2nd ed. (Jackson, Wyo.: Northern Rockies Conservation Cooperative, 1999). This compilation annotates fifty-seven references from scientists, managers, policy specialists, conservationists, journalists, and advocates who wrote about ecosystem management in the Greater Yellowstone region from 1971 through 1998.

12. Greater Yellowstone Coalition, *Threats to Greater Yellowstone* (Bozeman: Greater Yellowstone Coalition, 1984). This document can be obtained at the Greater Yellowstone Coalition, Box 1874, Bozeman, Mont., 59715; Greater Yellowstone Coalition, *Greater Yellowstone Challenges* (Bozeman: Greater Yellowstone Coalition, 1984).

13. Greater Yellowstone Coalition, *Greater Yellowstone Challenges 1986: An Inventory of Management Issues and Development Projects in the Greater Yellowstone Ecosystem* (Bozeman: Greater Yellowstone Coalition, 1986). This document can be obtained at the Greater Yellowstone Coalition, Box 1874, Bozeman, Mont., 59715.

14. D. Glick, M. Carr, and A. Harting, *An Environmental Profile of the Greater Yellowstone Ecosystem* (Bozeman: Greater Yellowstone Coalition, 1991), 99. This document can be obtained from the Greater Yellowstone Coalition, Box 1874, Bozeman, Mont., 59715.

15. A. Harting, D. Glick, and C. Rawlins, *Sustaining Greater Yellowstone: A Blueprint for the Future* (Bozeman: Greater Yellowstone Coalition, 1994). This document can be obtained at the Greater Yellowstone Coalition, Box 1874, Bozeman, Mont., 59715. This volume is encyclopedic in listing methods that could be used to improve management and policy.

16. *Greater Yellowstone Report* is published several times a year by the Greater Yellowstone Coalition, Box 1874, Bozeman, Mont., 59715; www.greater yellowstone.org.

17. The Jackson Hole Conservation Alliance produces newsletters several times a year. These feature current problems in the organization's area of operations. Available from Jackson Hole Conservation Alliance, Box 2728, Jackson, Wyo., 83001; www.jhalliance.org.

18. A. Chase, *Playing God in Yellowstone: The Destruction of America's First National Park* (Boston: Atlantic Monthly Press, 1986).

19. R. P. Reading, T. W. Clark, and S. R. Kellert, "Attitudes and Knowledge of People Living in the Greater Yellowstone Ecosystem," *Society and Natural Resources* 7 (1994): 349–65.

20. R. B. Keiter, "Taking Account of the Ecosystem on the Public Domain: Law and Ecology in the Greater Yellowstone Region," *University of Colorado Law Review* 60 (1989): 923–1007; R. B. Keiter, *Keeping Faith with Nature* (New Haven: Yale University Press, 2003).

21. R. Keiter and M. Boyce, eds., *The Greater Yellowstone Ecosystem: Redefining America's Wilderness Heritage* (New Haven: Yale University Press, 1991).

22. T. Palmer, *The Snake River: Window to the West* (Washington, D.C.: Island Press, 1991), 279.

23. Clark and Minta, *Greater Yellowstone's Future.*

24. S. A. Primm and T. W. Clark, "The Greater Yellowstone Policy Debate: What Is the Policy Problem?" *Policy Sciences* 29 (1996): 137–166.

25. P. Schullery, *Searching for Yellowstone: Ecology and Wonder in the Last Wilderness* (Boston: Houghton Mifflin, 1997), 1, 202.

26. T. W. Clark, "Interdisciplinary Problem Solving: Next Steps in the Greater Yellowstone Ecosystem," *Policy Sciences* 32 (1999): 393–414.

27. A. J. Hansen and R. DeFries, "Ecological Mechanisms Linking Nature Reserves to Surrounding Lands," in A. W. Biel, ed., *Beyond the Arch: Community and Conservation in Greater Yellowstone and East Africa,* Proceedings of the 7th Biennial Scientific Conference on the Greater Yellowstone Ecosystem (Mammoth, Wyo.: Yellowstone Center for Resources, 2004), 115–25; and A. J. Hansen, R. Rasker, B. Maxwell, J. J. Rotella, J. D. Johnnson, A. Wright Parmenter, U. Langner, W. B. Cohen, R. O. L. Lawrence, and M. P. V. Kraska, "Ecological Causes and Consequences of Demographic Change in the New West," *BioScience* 52, no. 2 (2002): 151–62.

28. G. O'Gara, *What You See in Clear Water: Indians, Whites, and a Battle over Water in the American West* (New York: Vintage Books, 2000).

29. C. M. Cromley, "The Killing of Grizzly Bear 209: Identifying Norms for Grizzly Bear Management," in T. W. Clark, A. R. Willard, and C. M. Cromley, eds., 2000, *Foundations of Natural Resources Policy and Management* (New Haven: Yale University Press, 2000), 173–220.

30. C. M. Cromley, "Bison Management in Greater Yellowstone," in R. D. Brunner, C. H. Colburn, C. M. Cromley, and R. A. Klein, eds., *Finding Common Ground: Governance and Natural Resources in the American West* (New Haven: Yale University Press, 2002), 126–58. See also R. D. Brunner, "Problems in Governance," in *Finding Common Ground,* 1–47.

31. T. W. Clark, D. Casey, and A. Halverson, eds., *Developing Sustainable Management Policy for the National Elk Refuge, Wyoming,* Bulletin No. 104 (New Haven: Yale School of Forestry and Environmental Studies, 2000), www.yale.edu/environmental/publications.

32. M. B. Rutherford, "Conceptual Frames, Values of Nature, and Key Symbols of Ecosystem Management: Landscape Scale Assessment in the Bridger-Teton National Forest" (Ph.D. diss., Yale School of Forestry and Environmental Studies, 2003).

33. G. Fergurson, *Hawk's Rest: A Season in the Remote Heart of Yellowstone* (Washington, D.C.: National Geographic Adventure Press, 2003).

34. U.S. Geological Survey, *Proposal: Linking Science and Management in the Greater Yellowstone Area* (Fort Collins, Colo.: Mid-Continent Ecological Science Center, Biological Resources Division, 1997), 2.

35. Greater Yellowstone Coordinating Committee, *Meeting Notes* (March 1997, Idaho Falls, Idaho) (Bozeman: Greater Yellowstone Coordinating Committee, 1997). This document can be obtained from the executive coordinator, GYCC, c/o U.S. Forest Service, Bozeman, Mont., 59715.

36. Greater Yellowstone Coordinating Committee, *Meeting Notes* (April 4–5, 2001) (Bozeman: Greater Yellowstone Coordinating Committee). This document can be obtained from the executive coordinator, GYCC, c/o U.S. Forest Service, Bozeman, Mont., 59715

37. Greater Yellowstone Coordinating Committee, "Where We Are Going . . . Current Issues and Priorities," *Greater Yellowstone Coordinating Committee Briefing Guide* (Bozeman: Greater Yellowstone Coordinating Committee, 2001), 14.

38. Yellowstone National Park, *Resources and Issues Handbook 2000* (Mammoth, Wyo.: Yellowstone National Park, Division of Interpretation, 2000).

39. Yellowstone National Park, *Resources and Issues Handbook 2001* (Mammoth, Wyo.: Yellowstone National Park, Division of Interpretation, 2001).

40. General Accounting Office, *Park Service: Managing for Results Could Strengthen Accountability,* Report to Congressional Requesters, GAO/RCED-97–125 (1997), 1–26; and General Accounting Office, *Forest Service Decision-Making: A Framework for Improving Performance,* GAO/RCED-97–71 (1997), 1–146.

41. Schullery, *Searching for Yellowstone,* 202.

42. Personal notes from the meeting.

43. J. A. Pritchard, *Preserving Yellowstone's Natural Conditions: Science and the Perception of Nature* (Lincoln: University of Nebraska Press, 1999), 294.

44. P. Lichtman and T. W. Clark, "Rethinking the 'Vision' Exercise in the Greater Yellowstone Ecosystem," *Society and Natural Resources* 7 (1994): 459–78.

45. E. Lewis, "Conspiracy Destroyed Greater Yellowstone Vision," *Greater Yellowstone Report* 10 (Winter 1993): 3.

46. Keiter and Boyce, *Greater Yellowstone Ecosystem,* 395, 409.

47. Personal notes, statement by Mike Finley, superintendent of Yellowstone National Park and chairperson of Greater Yellowstone Coordinating Committee at the time.

48. For an introduction to the difference between conventional and functional standpoints and thinking, see T. W. Clark, *The Policy Process: A Practical Guide for Natural Resources Professionals* (New Haven: Yale University Press, 2002), 123–25.

49. This model is an inclusive, comprehensive theory about humans in social process, decision making, and community life. It is also an analytic framework for researching the theory regardless of the context. Finally, the model contains propositions about how social process actually operates. This theory, which is a guide to functional analysis, is described in H. D. Lasswell, *A Pre-view of Policy Sciences* (New York: American Elsevier, 1971). For a more in-depth description and illustration of this model, see Lasswell and McDougal, *Jurisprudence for a Free Society;* and Clark et al., *Foundations of Natural Resources Policy and Management.*

50. See S. Primm and K. Murray, "Grizzly Bear Recovery: Living with Success," in Clark et al., *Coexisting with Large Carnivores.*

51. Examples are found in G. McLaughlin, S. A. Primm, and M. B. Rutherford, "Participatory Projects for Coexistence: Rebuilding Civil Society," and T. W. Clark and M.B. Rutherford, "The Institutional System of Wildlife Management: Making It More Effective," both in Clark et al., *Coexisting with Large Carnivores.*

52. Lasswell and McDougal, *Jurisprudence for a Free Society.*

53. L. Chen, *An Introduction to Contemporary International Law: A Policy-Oriented Perspective* (New Haven: Yale University Press, 1989), 16.

54. Lasswell and McDougal, *Jurisprudence for a Free Society.*

55. Ibid.

56. E. Deliso and T. W. Clark, *Finding Effective Elk Management Policy: A Contested Landscape of People, Wildlife, Livestock* (2007, in manuscript).

57. See W. H. Ulfelder, S. V. Poats, J. Recharte, and B. L. Dugelby, *Participatory Conservation: Lessons of the PALOMAP Study in Ecuador's Cayambe-Coca Ecological Reserve*, American Verde Working Paper Series No. 1b (Arlington, Va.: Nature Conservancy, 1997).

58. Clark et al., *Developing Sustainable Management Policy.*

59. G. D. Brewer and P. deLeon, *Foundations of Policy Analysis* (Homewood, Ill.: Dorsey Press, 1983); see also Clark et al., *Species and Ecosystem Conservation: An Interdisciplinary Approach*, Bulletin No. 105 (New Haven: Yale School of Forestry and Environmental Studies, 2001).

60. T. W. Clark, J. Tuxill, and M. S. Ashton, eds., "Appraising AMISCONDE at La Amistad Biosphere Reserve, Costa Rica: Finding Effective Conservation and Development," *Journal of Sustainable Forestry* 16, no. ½ (2003): 1–211; T. W. Clark, K. Ziegelmayer, M. Ashton, and Q. Newcomer, eds., "Conservation and Development in the Condor Bioreserve, Ecuador," *Journal of Sustainable Forestry* 17 (2004); T. W. Clark, M. Ashton, L. Dixon, and B. Petit, eds., "Conservation and Development in La Amistad: The Bocas del Toro (Panama) and Talamanca (Costa Rica) Experience," *Journal of Sustainable Forestry* 22, no. ½ (2006).

61. M. S. McDougal, and W. M. Reisman, eds., *International Law Essays: A Supplement of International Law in Contemporary Perspective* (Mineola, N.Y.: Foundation Press, 1981), 221.

62. Brunner et al., *Finding Common Ground*; R. D. Brunner and T. Steelman, *Adaptive Governance* (New York: Columbia University Press, 2005).

63. Clark, "Interdisciplinary Problem Solving."

64. This section is based on M. B. Rutherford and T. W. Clark, "Introduction," in T. W. Clark et al., *Management of the Greater Yellowstone Ecosystem: An Annotated Bibliography* (Jackson, Wyo.: Northern Rockies Conservation Cooperative, 1999), 1–15.

65. J. A. Weiss, "The Powers of Problem Definition: The Case of Government Paperwork," *Policy Sciences* 22 (1989): 97–121, quote on 97–98.

66. Greater Yellowstone Coalition, *Greater Yellowstone Challenges 1986: An Inventory of Management Issues and Development Projects in the Greater Yellowstone Ecosystem;* P. M. Hocker, "Yellowstone: The Region Is Greater Than the Sum of Its Parks," *Sierra* 64, no. 4 (1979): 8–12.

67. Conflicts are described in Chase, *Playing God in Yellowstone;* Hocker, "Yellowstone"; and R. Reese, *Greater Yellowstone: The National Park and Adjacent Wildlands*, Montana Geographic Series No. 6 (Helena, Mont.: Montana Magazine, 1984). P. C. Jobes, "The Greater Yellowstone Social System," *Conservation Biology* 5, no. 3 (1991): 387–94, describes the groups that exhibit different goals.

68. Brunner et al., *Finding Common Ground.*

69. Clark et al., *Foundations of Natural Resources Policy and Management.*

70. C. S. Holling, *Adaptive Environmental Assessment and Management* (New York: John Wiley and Sons, 1978); C. J. Walters, *Adaptive Management of Renewable Resources* (New York: Macmillan, 1986).

71. T. M. Power, "Ecosystem Preservation and the Economy in the Greater Yellowstone Area," *Conservation Biology* 5, no. 3 (1991): 395–404.

72. J. D. Johnson and D. J. Snepenger, "Application of the Tourism Life Cycle Concept in the Greater Yellowstone Region," *Society and Natural Resources* 6 (1993): 127–48.

73. C. E. Little, "The Challenge of Greater Yellowstone," *Wilderness* 51, no. 179 (1987): 18–56.

74. Reading et al., "Attitudes and Knowledge."

75. Hocker, "Yellowstone"; J. Berger, "Greater Yellowstone's Native Ungulates: Myths and Realities," *Conservation Biology* 5, no. 3 (1991): 353–63; T. W. Clark, "The Greater Yellowstone Ecosystem Idea: A Ten-Year Perspective," *NRCC (Northern Rockies Conservation Cooperative) News* 6 (1993): 17–20.

76. D. J. Mattson and M. M. Reid, "Conservation of the Yellowstone Grizzly Bear," *Conservation Biology* 5, no. 3 (1991): 364–72; D. J. Mattson, S. Herrero, R. G. Wright, and C. M. Pease, "Science and Management of Rocky Mountain Grizzly Bears," *Conservation Biology* 10, no. 4 (1996): 1013–25.

77. T. W. Clark, A. H. Harvey, R. D. Dorn, D. L. Genter, and C. Groves, *Rare, Sensitive, and Threatened Species of the Greater Yellowstone Ecosystem* (Jackson, Wyo.: Northern Rockies Conservation Cooperative; Montana Natural Heritage Program; The Nature Conservancy of Idaho, Montana, and Wyoming; Mountain West Environmental Services, 1989).

78. W. D. Newmark, "Legal and Biotic Boundaries of Western North American National Parks: A Problem of Congruence," *Biological Conservation* 33 (1985): 197–208; T. W. Clark and D. Zaunbrecher, "The Greater Yellowstone Ecosystem: The Ecosystem Concept in Natural Resource Policy and Management," *Renewable Resources Journal* (Summer 1987): 8–19; and S. A. Primm, "A Pragmatic Approach to Grizzly Bear Conservation," *Conservation Biology* 10, no. 4 (1996): 1026–35.

79. Keiter and Boyce, eds. *Greater Yellowstone Ecosystem;* Schullery, *Searching for Yellowstone.*

80. Berger, "Greater Yellowstone's Native Ungulates," 353–63. Schullery, *Searching for Yellowstone.*

81. Keiter and Boyce, *Greater Yellowstone Ecosystem;* and Schullery, *Searching for Yellowstone.*

82. Clark and Zaunbrecher, "Greater Yellowstone Ecosystem."

83. R. D. Barbee and J. D. Varley, "The Paradox of Repeating Error: Yellowstone National Park from 1872 to Biosphere Reserve and Beyond" (paper presented at the Conference for Managers of Biosphere Reserves, Great Smoky Mountains National Park, 1984), 27–29.

84. M. L. Corn, R. W. Gorte, and G. Siehl, "Issues Surrounding the Greater Yellowstone Ecosystem: A Brief Review, October 17, 1985," pp. 337–72 in U.S. Congress, House Committee on Interior and Insular Affairs, Subcommittee on Public Lands, *Greater Yellowstone Ecosystem,* oversight hearing before the Subcommittee on Public Lands and the Subcommittee on National Parks and Recreation of the Committee on Interior and Insular Affairs, House of Representatives, 99th Cong., 1st sess., hearing held in Washington, D.C., October 24, 1985 (Washington, D.C.: U.S. Government Printing Office, 1986); M. L. Corn and R. W. Gorte, "Yellowstone: Ecosystem, Resources, and Management," pp. 17–210 in *Greater Yellowstone Ecosystem: An Analysis of Data Submitted by Federal and State Agencies* (Washington, D.C.: U.S. Government Printing Office, 1987).

85. Sierra Club, *Yellowstone Under Siege: Oil and Gas Leasing in the Greater Yellowstone Region* (Washington, D.C.: Sierra Club, 1986).

86. R. Rasker, N. Tirrell, and D. Kloepfer, *The Wealth of Nature: New Economic Realities in the Yellowstone Region* (Washington, D.C.: Wilderness Society, 1992); and R. Rasker, "Rural Development, Conservation, and Public Policy in the Greater Yellowstone Ecosystem," *Society and Natural Resources* 6 (1993): 109–26.

87. Greater Yellowstone Coordinating Committee, *Vision for the Future: A Framework for Coordination in the Greater Yellowstone Area* (Billings, Mont.: Greater Yellowstone Coordinating Committee, 1990); R. M. Peterson, "The Greater Yellowstone Area—A Time for Coordinated Management" (speech before the Greater Yellowstone Coalition Annual Meeting, Lake Hotel, Yellowstone National

Park, Wyo., May 31, 1986); R. S. Gale, "Learning from the Past, Preparing for the Future," *Greater Yellowstone Report* 4, no. 3 (1987): 12–14; and T. McNamee, *Nature First: Keeping Our Wild Places and Wild Creatures Wild* (Boulder: Roberts Rinehart, 1987).

88. Lichtman and Clark, "Rethinking the 'Vision' Exercise."

89. R. L. Knight and P. B. Landres, eds., *Stewardship Across Boundaries* (Washington, D.C.: Island Press, 1998)

90. C. M. Lummis, ed., *Wyoming, Like No Place on Earth: Ways to Conserve Wyoming's Wonderful Open Lands: A Guide Book* (Cheyenne: Wyoming Governor's Office, 1997), 7.

91. R. Rasker and B. Alexander, *The New Challenge: People, Commerce and the Environment in the Yellowstone to Yukon Region* (Washington, D.C.: Wilderness Society, 1998).

92. K. N. Lee, *Compass and Gyroscope: Integrating Science and Politics for the Environment* (Washington, D.C.: Island Press, 1993).

93. S. L. Yaffee, *The Wisdom of the Spotted Owl: Policy Lessons for a New Century* (Washington, D.C.: Island Press, 1994).

94. W. H. Romme, in Clark and Minta, *Greater Yellowstone's Future,* 138.

95. See Hocker, "Yellowstone," 8–12.

96. Lasswell and McDougal, "The Social Process as a Whole," in *Jurisprudence for a Free Society,* 335–73; D. Torgerson, "Contextual Orientation in Policy Analysis: The Contribution of Harold D. Lasswell," *Policy Sciences* 18 (1985): 241–61; see also V. C. Arnspiger, *Personality in Social Process: Values and Strategies of Individuals in a Free Society* (Chicago: Follett, 1961).

Chapter 3. Leaders—Problem Solving

1. T. W. Clark, M. J. Stevenson, K. Ziegelmayer, and M. B. Rutherford, "Introduction: Leadership in Species and Ecosystem Conservation," in T. W. Clark, M. J. Stevenson, K. Ziegelmayer, and M. B. Rutherford, eds., *Species and Ecosystem Conservation: An Interdisciplinary Approach,* Bulletin No. 105 (New Haven: Yale School of Forestry and Environmental Studies, 2001), 9–15.

2. See J. S. Hammond, R. L. Keeney, and H. Raiffa, *Smart Choices: A Practical Guide to Making Better Decisions* (Boston: Harvard Business School Press, 1999).

3. H. D. Lasswell and M. S. McDougal, *Jurisprudence for a Free Society: Studies in Law, Science, and Policy* (New Haven: New Haven Press, 1992).

4. J. M. Burns, *Leadership* (New York: Harper and Row, 1978).

5. R. L. Daft, *Organizational Theory and Design* (St. Paul, Minn.: West Publishing, 1992).

6. B. Twight, *Organizational Values and Political Power: The Forest Service versus the Olympic National Park* (University Park: Pennsylvania State University Press, 1983), 25.

7. Burns, *Leadership;* G. B. Northcraft and M. A. Neal, *Organizational Behavior: A Management Challenge* (Chicago: Dryden Press, 1990).

8. D. Garber, *History of the Greater Yellowstone Coordinating Committee* (Bozeman: Greater Yellowstone Coordinating Committee, 1994). This document can be obtained from the executive coordinator, GYCC, c/o U.S. Forest Service, Bozeman, Mont., 59715; J. Lerum, *Historical Overview of Greater Yellowstone Coordinating Committee* (Bozeman: Greater Yellowstone Coordinating Committee, 1994). This document can be obtained from the executive coordinator, GYCC, c/o U.S. Forest Service, Bozeman, Mont., 59715.

9. Personal notes from the GYCC meeting, Jackson, Wyo., October 2, 1996.

10. Greater Yellowstone Coordinating Committee, GYCC Meeting Agenda (Billings, Mont.: Greater Yellowstone Coordinating Committee, November 3, 2004). This document can be obtained from the executive coordinator, GYCC, c/o U.S. Forest Service, Bozeman, Mont., 59715.

11. M. L. Corn, R. W. Gorte, and G. Siehl, "Issues Surrounding the Greater Yellowstone Ecosystem: A Brief Review, October 17, 1985," pp. 337–72 in U.S. Congress, House Committee on Interior and Insular Affairs, Subcommittee on Public Lands, *Greater Yellowstone Ecosystem,* oversight hearing before the Subcommittee on Public Lands and the Subcommittee on National Parks and Recreation of the Committee on Interior and Insular Affairs, House of Representatives, 99th Cong., 1st sess., hearing held in Washington, D.C., October 24, 1985 (Washington, D.C.: U.S. Government Printing Office, 1986); M. L. Corn and R. W. Gorte, "Yellowstone: Ecosystem, Resources, and Management," pp. 17–210 in *Greater Yellowstone Ecosystem: An Analysis of Data Submitted by Federal and State Agencies* (Washington, D.C.: U.S. Government Printing Office, 1987).

12. Corn et al., "Issues Surrounding the Greater Yellowstone Ecosystem"; and Corn and Gorte, "Yellowstone."

13. Garber, *History of the Greater Yellowstone Coordinating Committee,* 2.

14. Ibid., 1.

15. Greater Yellowstone Coordinating Committee, *The Greater Yellowstone Area: An Aggregation of National Park and National Forest Management Plans* (Yellowstone National Park, Wyo.: USDA Forest Service and USDI National Park Service, 1987).

16. Personal notes from the GYCC meeting, Jackson, Wyo., October 2, 1996.

17. Greater Yellowstone Coordinating Committee, *Vision for the Future: A Framework for Coordination in the Greater Yellowstone Area* (Billings, Mont.: Greater Yellowstone Coordinating Committee, 1990).

18. P. Lichtman and T. W. Clark, "Rethinking the 'Vision' Exercise in the Greater Yellowstone Ecosystem," *Society and Natural Resources* 7 (1994): 459–78.

19. Garber, *History of the Greater Yellowstone Coordinating Committee,* 1.

20. Personal notes from the GYCC meeting, Jackson, Wyo., October 2–3, 1996.

21. Garber, *History of the Greater Yellowstone Coordinating Committee,* 2.

22. Personal notes from the GYCC meeting, Jackson, Wyo., October 2–3, 1996.

23. Garber, *History of the Greater Yellowstone Coordinating Committee,* 2.

24. Ibid.; U.S. General Accounting Office, *Ecosystem Management: Additional Actions Needed to Adequately Test a Promising Approach* (Washington, D.C.: U.S. General Accounting Office, 1994).

25. Garber, *History of the Greater Yellowstone Coordinating Committee,* 3.

26. Greater Yellowstone Coordinating Committee, *Agenda Concepts* (Billings, Mont.: Greater Yellowstone Coordinating Committee, October 2–3, 1996). This document can be obtained from the executive coordinator, GYCC, c/o U.S. Forest Service, Bozeman, Mont., 59715; Greater Yellowstone Coordinating Committee, *Proposal for a National Spatial Data Infrastructure Information Center and Sharing of Geographic Information Systems Technology Among Local, State, Federal Governments within the Greater Yellowstone Area* (Billings, Mont.: Greater Yellowstone Coordinating Committee, November 3, 2004). This document can be obtained from the executive coordinator, GYCC, c/o U.S. Forest Service, Bozeman, Mont., 59715.

27. Personal notes from the GYCC meeting, Jackson, Wyo., October 2–3, 1996.

28. Ibid.

29. Greater Yellowstone Coordinating Committee, *Position Description: Executive Coordinator GS-301–12/13: Position Number* (Billings, Mont.: Greater Yellowstone Coordinating Committee, 2000).

30. Greater Yellowstone Coordinating Committee, *Greater Yellowstone Coordinating Committee Briefing Guide* (Bozeman: Greater Yellowstone Coordinating Committee, 2001), 10.

31. Personal notes from the GYCC meeting, Jackson, Wyo., December 6–7, 2000.

32. Greater Yellowstone Coordinating Committee, *Briefing Guide.*

33. Ibid., 2.

34. U.S. Government, Memorandum of Understanding between the Rocky Mountain Region National Park Service, the Northern, Rocky Mountain and Intermountain Region Forest Service, and the Mountain Prairie Region Fish and Wildlife Service (2003). This document can be obtained from the executive coordinator, GYCC, c/o U.S. Forest Service, Bozeman, Mont., 59715.

35. Daft, *Organizational Theory and Design.*

36. United Nations Development Programme (1995), 14, cited in H. J. Aslin, N. A. Mazur, and A. L. Curtis, 2002, *Identifying Regional Skill and Training Needs for Integrated Natural Resource Management Planning* (Canberra, Australia: Bureau of Rural Sciences, 1995), 11.

37. Memorandum of Understanding between the Rocky Mountain Region, National Park Service and the Northern, Rocky Mountain, and Intermountain Regions, Forest Service, 1987, Appendix A in the Greater Yellowstone Area Aggregation, pp. 5.2–5.3.

38. Greater Yellowstone Coordinating Committee, Meeting Minutes (Billings, Mont.: Greater Yellowstone Coordinating Committee, 1986). See discussion of the 1964 Memorandum of Understanding among the agencies.

39. Greater Yellowstone Coordinating Committee, *Strategic Thinking—Adjusting the Course,* Session Documentation (Jackson, Wyo.: Greater Yellowstone Coordinating Committee, April 2–3, 2003), 2. This document can be obtained from the executive coordinator, GYCC, c/o U.S. Forest Service, Bozeman, Mont., 59715.

40. The federal agencies operate under various federal laws, including but not limited to the following: 1916 National Park Service Act; 1939 Fish and Wildlife Service Act; 1976 National Forest Management Act; 1973 Endangered Species Act; and Multiple Use, Sustained Yield Act. These and other laws can be found in R. V. Percival, A. S. Miller, C. H. Schroeder, and J. P. Leape, *Environmental Regulation: Law, Science, and Policy,* 3rd ed. (New York: Aspen Law and Business, 2000), 937–38, 973–74.

41. P. Schullery and L. Whittlesey, *Myth and History in the Creation of Yellowstone National Park* (Lincoln: University of Nebraska Press, 2003).

42. A. S. Leopold, S. A. Cain, C. Cottam, I. N. Gabrielson, and T. L. Kimball, "Wildlife Management in the National Parks," *Transactions of the North American Wildlife Conference* 28 (1963): 27–45; see also D. Despain, D. Houston, M. Meagher, and P. Schullery, *Wildlife in Transition: Man and Nature on Yellowstone's Northern Range* (Boulder: Roberts Rhinehart, 1986).

43. R. W. Sellars, *Preserving Nature in the National Parks: A History* (New Haven: Yale University Press, 1997).

44. D. A. Schön, *Beyond the Stable State* (New York: W. W. Norton, 1971).

45. J. N. Clarke and D. McCool, *Staking Out the Terrain: Power Differentials among Natural Resource Management Agencies* (Albany: State University of New York Press, 1985); and R. B. Keiter, *Keeping Faith with Nature* (New Haven: Yale University Press, 2003).

46. Daft, *Organizational Theory and Design,* 86–87.

47. D. Taylor and T. W. Clark, "Management Context: People, Animals, and Institutions," and T. W. Clark and M. B. Rutherford, "The Institutional System of

Wildlife Management: Making It More Effective," in T. W. Clark, M. B. Rutherford, and D. Casey, eds., *Coexisting with Large Carnivores: Lessons from Greater Yellowstone* (Washington, D.C.: Island Press, 2005); B. Stanford, *Myths and Modern Man* (New York: Pocket Books, 1972); and H. A. Murray, ed., *Myth and Mythmaking* (Boston: Beacon Press, 1960).

48. T. W. Clark, personal communication with Larry Timchak (Billings, Mont.: October, 8, 1997.

49. J. L. Sax and R. B. Keiter, "Glacier National Park and Its Neighbors: A Study of Federal Interagency Relations," *Ecology Law Review* 14, no. 2 (1987): 207–63.

50. See J. G. March and H. A. Simon, *Organizations* (New York: John Wiley and Sons, 1958); see R. M. Cyert and J. G. March, *A Behavioral Theory of the Firm* (Englewood Cliffs, N.J.: Prentice-Hall, 1963); and J. D. Thompson, 1967, *Organizations in Action* (New York: McGraw-Hill, 1967).

51. Problems are interesting entities. See J. A. Weiss, "The Powers of Problem Definition: The Case of Government Paperwork," *Policy Sciences* 22 (1989): 97–121; D. Dery, *Problem Definition in Policy Analysis* (Lawrence: University of Kansas Press, 1984); and A. Miller, *Environmental Problem Solving: Psychosocial Barriers to Adaptive Management* (New York: Springer, 1999). For applications in the Yellowstone region, see S. A. Primm and T. W. Clark, "The Greater Yellowstone Policy Debate: What Is the Policy Problem?" *Policy Sciences* 29 (1996): 137–65; and T. W. Clark, A. P. Curlee, and R. P. Reading, "Crafting Effective Solutions to the Large Carnivore Conservation Problem," *Conservation Biology* 10 (1996): 940–48.

52. See T. W. Clark, "Interdisciplinary Problem Solving in Endangered Species Conservation: The Yellowstone Grizzly Bear Case," in R. P. Reading and B. J. Miller, eds., *Endangered Animals* (Westport, Conn.: Greenwood, 2000), 285–301.

53. See J. Dewey, "How We Think," in J. Boydston, ed., *J. Dewey: The Middle Works 1899–1924,* vol. 6 (1910–1911) (Carbondale: Southern Illinois University Press, 1978), 177–356 (Mineola, N.Y.: Dover Publications, 1997).

54. For example, see T. W. Clark, "Interdisciplinary Problem Solving: Next Steps in the Greater Yellowstone Ecosystem," *Policy Sciences* 32 (1999): 393–414; and R. D. Brunner, "Context-Sensitive Monitoring and Evaluation for the World Bank," *Policy Sciences* 37 (2004): 103–36.

55. Lasswell and McDougal, *Jurisprudence for a Free Society.*

56. E. Roe, "Varieties of Issue Incompleteness and Coordination: An Example from Ecosystem Management," *Policy Sciences* 34 (2001): 111–34.

57. W. Ascher and R. Healy, *Natural Resource Policymaking in Developing Countries* (Durham: Duke University Press, 1990).

58. On policy orientation see H. D. Lasswell, *A Pre-view of Policy Sciences* (New York: American Elsevier, 1971); R. P. Wallace and T. W. Clark, "Solving Problems in Endangered Species Conservation: An Introduction to Problem Orientation," in R. P. Wallace, T. W. Clark, and R. P. Reading, eds., *An Interdisciplinary Approach to Endangered Species Recovery: Concepts, Applications, and Case,* special issue of *Endangered Species Update* 19, no. 4 (2002): 81–86.

59. G. D. Brewer and P. deLeon, *The Foundations of Policy Analysis* (Homewood, Ill.: Dorsey Press, 1983), 22.

Chapter 4. Leaders—Cooperation and Demonstrations

1. R. Daft, *Organizational Theory and Design.* See also B. S. Low, "Human Behavior and Conservation," *Endangered Species Update* 21, no. 1 (2004): 14–22; R. E. Kranton, "The Formation of Cooperative Relationships," *Journal of Law, Economics,*

and Organization 12, no. 1 (1996): 214–33; and R. E. Kranton, "Reciprocal Exchange: A Self Sustaining System," *American Economic Review* 86, no. 4 (1996): 830–49.

2. See M. Thompson and M. Warburton, "Knowing Where to Hit It: A Conceptual Framework for the Sustainable Development of the Himalaya," *Mountain Research and Development* 5, no. 3 (1985): 203–20; H. Cortner and M. Moote, *The Politics of Ecosystem Management* (Washington, D.C.: Island Press, 1999); and T. W. Clark, "Interdisciplinary Problem Solving: Next Steps in the Greater Yellowstone Ecosystem," *Policy Sciences* 32 (1999): 393–414.

3. See D. C. Zbicz, "Transboundary Cooperation in Conservation: A Global Survey of Factors Influencing Cooperation between Internationally Adjoining Protected Areas" (Ph.D. diss., Durham, N.C.: Duke University, 1999); and D. C. Zbicz, "Imposing Transboundary Conservation: Cooperation between Internationally Adjoining Protected Areas," *Journal of Sustainable Forestry* 17, no. ½ (2003): 21–37.

4. For example, see L. S. Etheredge, *Can Governments Learn?* (New York: Pergamon Press, 1985); P. R. Leeuw, R. C. Rist, and R. C. Sonnichsen, *Can Governments Learn? Comparative Perspectives on Evaluation and Organizational Learning* (New Brunswick, N.J.: Transaction Publishers, 1994).

5. R. D. Putnam, *Bowling Alone: The Collapse and Revival of American Community* (New York: Simon and Schuster, 2000); J. D. Montgomery and A. Inkeles, eds., "Social Capital as a Policy Resource," *Policy Sciences* 33, nos. 3–4 (2000): 227–494; J. S. Coleman, *Power and the Structure of Society* (New York: W. W. Norton, 1974).

6. Daft, *Organizational Theory and Design,* 299.

7. Putnam, *Bowling Alone,* 3. See also E. M. Uslaner, "Democracy and Social Capital," in M. E. Warren, ed., *Democracy and Trust* (Cambridge: Cambridge University Press, 1999), 121–150.

8. T. P. Duane, "Community Participation in Ecosystem Management," *Ecology Law Quarterly* 24, no. 4 (1997): 775.

9. C. M. Cromley, "The Killing of Grizzly Bear 209: Identifying Norms for Grizzly Bear Management," in T. W. Clark, A. R. Willard, and C. M. Cromley, eds., *Foundations of Natural Resources Policy and Management* (New Haven: Yale University Press, 2000), 173–220; C. M. Cromley, "Bison Management in Greater Yellowstone," in R. D. Brunner, C. H. Colburn, C. M. Cromley, and R. A. Klein, eds., *Finding Common Ground: Governance and Natural Resources in the American West* (New Haven: Yale University Press, 2002), 126–158.

10. J. Padwe, "Participatory Conservation in the Condor Bioreserve, Ecuador: Representations, Decision Processes, and Underlying Assumptions," *Journal of Sustainable Forestry* 18, no. ⅔ (2004): 107–37; M. P. Pimbert and J. N. Pretty, *Parks, People, and Professionals: Putting Participation into Social Development,* United Nations Research Institute for Social Development Discussion Paper #57 (Geneva, Switz.: UNRISD, 1995).

11. M. S. McDougal, H. D. Lasswell, and L. Chen, *Human Rights and World Public Order: The Basic Polices of an International Law of Human Dignity* (New Haven: Yale University Press, 1980).

12. J. S. Dryzek, *Discursive Democracy: Politics, Policy, and Political Science* (New York: Cambridge University Press, 1990); W. A. Shutkin, *The Land That Could Be: Environmentalism and Democracy in the Twenty-First Century* (Cambridge: MIT Press, 2000); and B. A. Minteer and R. E. Manning, eds., *Reconstructing Conservation: Finding Common Ground* (Washington, D.C.: Island Press, 2003). See also G. McLaughlin, S. A. Primm, and M. B. Rutherford, "Participatory Projects for Coexistence: Rebuilding Civil Society" in T. W. Clark, M.B. Rutherford, and D. Casey,

eds., *Coexisting with Large Carnivores: Lessons from Greater Yellowstone* (Washington, D.C.: Island Press, 2005), 224–65; and J. Bohman, *Public Deliberation: Pluralism, Complexity, and Democracy* (Cambridge: MIT Press, 1996).

13. T. P. Duane, "Community Participation in Ecosystem Management," *Ecology Law Quarterly* 24, no. 4 (1997): 780.

14. Montgomery and Inkeles, "Social Capital as a Policy Resource."

15. G. D. Brewer, "The Policy Process as a Perspective for Understanding," in E. Zigler, S. O. L. Kagan, and E. Klugman, eds., *Children, Families, and Government* (New York: Cambridge University Press, 1983), 57–76. Also see H. D. Lasswell, *The Decision Process: Seven Categories of Functional Analysis* (College Park: Bureau of Government Research, University of Maryland, 1956); W. M. Reisman, "Institutions and Practices for Restoring and Maintaining Public Order," *Duke Journal of Comparative and International Law* 6, no. 1 (1995): 175; and W. Ascher and R. Healy, "Knowledge in the Policy Process," *Policy Sciences* 28 (1995): 1–19.

16. More detail on this model is in G. D. Brewer and P. deLeon, *Foundations of Policy Analysis* (Homewood, Ill.: Dorsey Press, 1983).

17. See Greater Yellowstone Coordinating Committee, Agenda Concepts (Jackson, Wyo.: Greater Yellowstone Coordinating Committee, Fall 1996); and Greater Yellowstone Coordinating Committee, Fall Meeting Agenda (Jackson, Wyo.: Greater Yellowstone Coordinating Committee, October 2–3, 1996). These documents can be obtained from the executive coordinator, GYCC, c/o U.S. Forest Service, Bozeman, Mont., 59715.

18. See Greater Yellowstone Coordinating Committee, GYCC Fall Meeting Notes (Jackson, Wyo.: Greater Yellowstone Coordinating Committee, October 8–9, 1997); Greater Yellowstone Coordinating Committee, GYCC Meeting Agenda (Jackson, Wyo.: Greater Yellowstone Coordinating Committee, October 8–9, 1997). These documents can be obtained from the executive coordinator, GYCC, c/o U.S. Forest Service, Bozeman, Mont., 59715

19. H. D. Lasswell, *A Pre-view of Policy Sciences* (New York: American Elsevier, 1971).

20. Ibid., 85–97.

21. This list of common weaknesses comes from W. Ascher and R. Healy, *Natural Resource Policymaking in Developing Countries* (Durham: Duke University Press, 1990), 159–96; and T. W. Clark, *Averting Extinction: Reconstructing Endangered Species Recovery* (New Haven: Yale University Press, 1997), 171.

22. P. Lichtman and T. W. Clark, "Rethinking the 'Vision' Exercise in the Greater Yellowstone Ecosystem," *Society and Natural Resources* 7 (1994): 459–78.

23. U.S. Congress, House Committee on Interior and Insular Affairs, Subcommittee on Public Lands, *Greater Yellowstone Ecosystem,* oversight hearing before the Subcommittee on Public Lands and the Subcommittee on National Parks and Recreation of the Committee on Interior and Insular Affairs, House of Representatives, 99th Cong., 1st sess., hearing held in Washington, D.C., October 24, 1985 (Washington, D.C.: U.S. Government Printing Office, 1986); *Greater Yellowstone Ecosystem: An Analysis of Data Submitted by Federal and State Agencies* (Washington: U.S. Government Printing Office, 1987).

24. J. Berry, G. D. Brewer, J. C. Gordon, and D. R. Patton, "Closing the Gap between Ecosystem Management and Ecosystem Research," *Policy Sciences* 31 (1998): 55–80. Also see G. D. Brewer, "Science to Serve the Common Good," *Environment* 39 (1997): 25–28.

25. C. E. Lindblom, "The Science of 'Muddling Through,' " *Public Administration Review* 19 (1954): 79–88; M. D. Cohen, J. G. March, and J. P. Olsen, "A Garbage

Can Model of Organizational Choice," *Administrative Science Quarterly* 17 (1972): 1–25; M. D. Cohen and J. G. March, *Leadership and Ambiguity: The American College President* (New York: McGraw-Hill, 1974). Incrementalistic strategies appear "indecisive, makeshift, timid, narrow, inconclusive, and procrastinating." See C. E. Lindblom and E. J. Woodhouse, *The Policy Making Process* (Englewood Cliffs, N.J.: Prentice-Hall, 1993), 32.

26. E. B. Hass, *When Knowledge Is Power: Three Models of Change in International Organizations* (Berkeley: University of California Press, 1990).

27. See Cortner and Moote, *Politics of Ecosystem Management;* Clark, *Policy Sciences* 32, 393–414; and Thompson and Warburton, "Knowing Where to Hit It."

28. Greater Yellowstone Coordinating Committee, "GYCC at Work," in *Greater Yellowstone Coordinating Committee Briefing Guide* (Billings, Mont.: Greater Yellowstone Coordinating Committee, 2001), 9.

29. Personal conversation with Steve Mealey in May 1984.

30. See B. Goldstein, "The Struggle over Ecosystem Management at Yellowstone," *BioScience* 42 (1992): 183–87; and A. J. Hanson, R. Rasker, B. Maxwell, J. J. Rotella, J. D. Johnson, A. W. Parmenter, U. Langer, W. B. Cohen, R. L. Lawrence, and M. P. V. Kraska, "Ecological Causes and Consequences of Demographic Change in the New West," *BioScience* 52, no. 2 (2002): 151–62.

31. Greater Yellowstone Coordinating Committee, *Yellowstone Area: An Aggregation of National Park and National Forest Management Plans* (Billings, Mont.: Greater Yellowstone Coordinating Committee, 1987).

32. See T. W. Clark, "An Informational Approach to Sustainability: 'Intelligence' in Conservation and Natural Resource Management Policy," *Journal of Sustainable Forestry,* in press.

33. L. Mintzmyer, "The Keys to the Treasure Chest" (remarks to the 1992 annual meeting of the Greater Yellowstone Coalition, West Yellowstone, Mont., May 29, 1991), 1–6.

34. Greater Yellowstone Coordinating Committee, *Vision for the Future: A Framework for Coordination in the Greater Yellowstone Area* (Billings, Mont.: Greater Yellowstone Coordinating Committee, 1990); and Greater Yellowstone Coordinating Committee, *A Framework for Coordination of National Parks and National Forests in the Greater Yellowstone Area* (Billings, Mont.: Greater Yellowstone Coordinating Committee, 1991).

35. Greater Yellowstone Coordinating Committee, *Framework for Coordination,* 1.

36. R. D. Barbee, P. Schullery, and J. D. Varley, "The Yellowstone Vision: An Experiment That Failed or a Vote for Posterity?" in R. Greenberg, ed., *Proceedings of a Conference on Partnerships in Parks and Preservation* (Washington, D.C.: National Park Service and National Park and Conservation Association, 1991), 80–85.

37. M. Milstein, "Yellowstone Plan Gutted, Critics Say," *Billings Gazette,* September 12, 1991, 1A.

38. E. Lewis, "Goodbye Vision, Mumma, Mintzmyer," *Greater Yellowstone Report* 8, no. 4 (1991): 3.

39. Lichtman and Clark, "Rethinking the 'Vision' Exercise."

40. Incident analysis from W. M. Reisman and A. R. Willard, eds., *International Incidents* (Princeton: Princeton University Press, 1988); Cromley, "Killing of Grizzly Bear 209."

41. Greater Yellowstone Coordinating Committee, *FY 2001 GYCC Project Fund Proposals* (December 13, 2000); *FY 2001 GYCC Project Funding* (March 26, 2001); *GYCC Project Funding* (December 19, 2001); *FY 2002 GYCC Project Fund Proposals*

(December 19, 2001); *FY 2002 GYCC Project Fund Proposals* (March 20, 2002); *Summary of GYCC Funded Projects, FY 2000–2002* (July, 2002); *Project Funding Process* (2003); *FY 2003 GYCC Projects Funded* (March 5, 2003); *FY 2003 GYCC Projects* (January 23, 2003); *FY 2004 GYCC Project Funding* (January 27, 2004), Greater Yellowstone Coordinating Committee, Bozeman, Mont. These documents can be obtained from the executive coordinator, GYCC, c/o U.S. Forest Service, Bozeman, Mont., 59715.

42. Greater Yellowstone Coordinating Committee, *Briefing Guide*, 10–13.

43. Greater Yellowstone Coordinating Committee, "Status of GYCC Committees," *Briefing Guide*, 7–8.

44. Haas, "Introduction: Epistemic Communities and International Environmental Cooperation."

Chapter 5. Overall Assessment—Leaders, Bureaucracy, and Context

1. A review is in W. Ascher and B. Hirschfelder-Ascher, *Revitalizing Political Psychology: The Legacy of Harold D. Lasswell* (Mahwah, N.J.: Lawrence Erlbaum Associates, 2005).

2. H. D. Lasswell and M. S. McDougal, "Political Personality," in *Jurisprudence for a Free Society: Studies in Law, Science, and Policy* (New Haven: New Haven Press, 1992). This section is based on this source and other writings by Lasswell.

3. H. D. Lasswell, *Democratic Character* (Glencoe, Ill.: Free Press, 1951).

4. V. C. Arnspiger, *Personality in Social Process: Values and Strategies of Individuals in a Free Society* (Chicago, Ill.: Follett, 1961); H. A. Simon, *Models of Man* (New York: John Wiley and Sons, 1957); and H. A. Simon, *Reason in Human Affairs* (Stanford: Stanford University Press, 1983).

5. For example, see Arbinger Institute, *Leadership and Self-Deception: Getting Out of the Box* (San Francisco: Berrett-Koehler, 2000).

6. C. E. Lindblom, "The Science of 'Muddling Through,' " *Public Administration Review* 19 (1959): 79–88.

7. A. Rapoport, *Fights, Games, and Debates* (Ann Arbor: University of Michigan Press, 1960).

8. C. Taylor, *Sources of the Self: The Making of the Modern Identity* (Cambridge: Harvard University Press, 1989); and D. Wahrman, *The Making of the Modern Self* (New Haven: Yale University Press, 2004).

9. D. Taylor and T. W. Clark, "Management Context: People, Animals, and Institutions," and T. W. Clark and M. B. Rutherford, "The Institutional System of Wildlife Management: Making It More Effective," in T. W. Clark, M. B. Rutherford, and D. Casey, eds., *Coexisting with Large Carnivores: Lessons from Greater Yellowstone* (Washington, D.C.: Island Press, 2005), 28–68, 211–53.

10. See A. Miller, "Cognitive Styles and Environmental Problem-Solving," *International Journal of Environmental Studies* 26 (1985): 21–31. Many authors talk about positivism and its limitations in management and policy. See, for example, G. Morocco, "Positivist Beliefs among Policy Professionals: An Empirical Investigation," *Policy Sciences* 34 (2001): 381–401, and see, L. Chen, *An Introduction to Contemporary International Law: A Policy-oriented Perspective* (New Haven: Yale University Press, 1989).

11. R. D. Brunner, C. H. Colburn, C. M. Cromley, R. A. Klein, and E. A. Olson, eds., *Finding Common Ground: Governance and Natural Resources in the American West* (New Haven: Yale University Press, 2002).

12. Ibid.

13. Chen, *Introduction to Contemporary International Law.*

14. I. M. Martin and T. A. Steelman, "Using Multiple Methods to Understand Agency Values and Objectives: Lessons for Public Lands Management," *Policy Sciences* 37 (2004): 37–69.

15. C. M. Ryan, "Leadership in Collaborative Policy-Making: An Analysis of Agency Roles in Regulatory Negotiations," *Policy Sciences* 34, nos. 3–4 (2001): 221–45; E. C. Poncelet, "Personal Transformation in Multistakeholder Environmental Partnerships," *Policy Sciences* 34, nos. 3–4 (2001): 273–301; and T. W. Clark, *Averting Extinction: Reconstructing Endangered Species Recovery* (New Haven: Yale University Press, 1997), 167–87.

16. H. D. Lasswell, *Psychopathology and Politics* (New York: Viking Press, 1960); and H. D. Lasswell, *World Politics and Personal Insecurity* (New York: Free Press, 1935).

17. Ascher and Hirschfelder-Ascher, *Revitalizing Political Psychology.*

18. R. Westrum, "A Typology of Organizational Cultures," *Quality and Safety in Health Care,* available at www.qshc.com; A. J. Adamski, and R. Westrum, "Requisite Imagination: The Fine Art of Anticipating What Might Go Wrong," in E. Hollnagel, ed., *Handbook of Cognitive Task Design* (Mahwah, N.J.: Lawrence Erlbaum Associates, 2003), 193–220; I. Smillie and J. Hailey, *Managing for Change: Leadership, Strategy, and Management* (Oxford: Oxford University Press, 2001); and N. Tichy, *Managing Strategic Change: Technical, Political, and Cultural Dynamics* (New York: John Wiley and Sons, 1983).

19. This section is based on Y. Dror, *The Capacity to Govern* (Portland, Ore.: Frank Cass, 1994).

20. T. W. Clark, M. J. Stevenson, K. Ziegelmayer, and M. B. Rutherford, eds., *Species and Ecosystem Conservation: An Interdisciplinary Approach,* Bulletin No. 105 (New Haven: Yale School of Forestry and Environmental Studies, 2001), 270.

21. M. S. McDougal, W. M. Riesman, and A. R. Willard, "The World Community: A Planetary Social Process," *University of California Law Review* 21 (1988): 807–972.

22. R. D. Brunner, "The Policy Movement as a Policy Problem," *Policy Sciences* 24 (1991): 65–98.

23. J. C. Scott, *Seeing Like a State: How Certain Schemes to Improve the Human Condition Have Failed* (New Haven: Yale University Press, 1998). See also A. Willard, "Scott, James C., *Seeing Like a State: How Certain Schemes to Improve the Human Condition Have Failed,*" book review, *Policy Sciences* 33, no. 1 (2000): 107–15.

24. There are many examples of resistance. See G. Reynolds, *Promise or Threat? A Study of "Greater Yellowstone Ecosystem" Management* (Riverton, Wyo.: WeCARE, 1987); and McDougal et al., "World Community." McDougal et al. (1988) took on mapping the entire world community as a single planetary social process spanning many centuries The globalization phenomena sweeping the world today makes this article that much more important. Another example of contextuality focuses on mapping logging in the American West and other cases (see T. W. Clark, *The Policy Process: A Practical Guide for Natural Resource Professionals* [New Haven: Yale University Press, 2002]).

25. See J. S. Dryzek, *The Politics of the Earth: Environmental Discourses* (Oxford: Oxford University Press, 1997); and S. Swaffield, "Contextual Meanings in Policy Discourse: A Case Study of Language Use Concerning Resource Policy in the New Zealand High Country," *Policy Sciences* 31 (1998): 199–224.

26. Personal notes from the GYCC meeting, Jackson, Wyo., December 6–7, 2000.

27. P. L. Berger and T. Luckmann, *Social Construction of Reality: A Treatise in the Sociology of Knowledge* (New York: Penguin, 1987).

28. J. A. Throgmorton, "The Rhetorics of Policy Analysis," *Policy Sciences* 24 (1991): 153.

29. J. P. Brosius, "What Counts as Local Knowledge in Global Environmental Assessments and Conventions?" (presentation at plenary session on "Integrating Local and Indigenous Perspectives into Assessments and Conventions" at a conference on Bridging Scales and Epistemologies: Linking Local Knowledge and Global Science in Multi-Scale Assessments, Bibliotheca Alexandrina, Alexandria, Egypt, March 17–20, 2004), and papers cited in this report.

30. J. Forester, *The Deliberative Practitioner: Encouraging Participatory Planning Processes* (Cambridge: MIT Press, 2000), 26.

31. Ibid., 11.

32. T. Sager, *Communicative Planning Theory* (Aldershot, N.Y.: Avebury Press, 1994).

33. C. W. Thomas, *Bureaucratic Landscapes: Interagency Cooperation and the Preservation of Biotic Diversity* (Cambridge: MIT Press, 2003).

34. M. Thompson and M. Warburton, "Knowing Where to Hit It: A Conceptual Framework for the Sustainable Development of the Himalaya," *Mountain Research and Development* 5, no. 3 (1985): 203–20.

35. J. Q. Wilson, *Bureaucracy: What Government Agencies Do and Why They Do It* (New York: BasicBooks, 1989).

36. G. T. Allison, *Essence of Decision: Explaining the Cuban Missile Crisis* (Boston: Little, Brown, 1971), 67.

37. C. E. Lindblom and E. J. Woodhouse, *The Policy-Making Process* (Englewood Cliffs, N.J.: Prentice-Hall, 1993).

38. D. Mosse, "Is Good Policy Unimplementable? Reflections on the Ethnography of Aid Policy and Practice," *Development and Change* 35, no. 4 (2004): 639–71, 667.

39. S. L. Yaffee, "Lessons about Leadership from the History of the Spotted Owl Controversy," *Natural Resources Journal* 35 (1995): 381–412, 402.

40. See J. L. Sax and R. B. Keiter, "Glacier National Park and Its Neighbors: A Study of Federal Interagency Relations, *Ecology Law Quarterly* 14 (1987): 207–63.

41. A. Flores and T. W. Clark, "Finding Common Ground in Biological Conservation: Beyond the Anthropocentric vs. Biocentric Controversy," in Clark et al., *Species and Ecosystem Conservation*, 241–52.

42. L. S. Etheredge, *A World of Men: The Private Sources of American Foreign Policy* (Cambridge: MIT Press, 1976); L. S. Etheredge, "Hardball Politics: A Model," *Political Psychology* 1, no. 1 (1979): 3–26; and L. S. Etheredge, *The Case of the Unreturned Cafeteria Trays* (Washington, D.C.: American Political Science Association, 1976).

43. Much has been written on organizational learning: C. Arygris, *On Organizational Learning* (Cambridge, Mass.: Blackwell, 1992); C. Arygris and D. Schön, *Organizational Learning* (Reading, Mass.: Addison-Wesley, 1978); I. L. Janis, *Victims of Groupthink: A Psychological Study of Foreign Policy Decisions and Fiascos* (Boston: Houghton Mifflin, 1972); G. Morgan, *Images of Organization* (Beverly Hills: Sage Publications, 1986); P. M. Senge, *The Fifth Discipline: The Art and Practice of the Learning Organization* (New York: Doubleday, 1990); D. Osborne and T. G. Gaebler, *Reinventing Government: How the Entrepreneurial Spirit is Transforming* (New York: Penguin Books, 1993); P. R. Leeuw, R. C. Rist, and R. C. Sonnichsen, *Can Governments Learn? Comparative Perspectives on Evaluation and Organizational Learning* (New Brunswick, N.J.: Transaction, 1994).

44. E. Roe, "Varieties of Issue Incompleteness and Coordination: An Example from Ecosystem Management," *Policy Sciences* 34 (2001): 111–34.

45. Ibid.

46. D. E. Stokes, *Pasteur's Quadrant: Basic Science and Technological Innovations* (Washington, D.C.: Brookings Institution Press, 1997); J. Wilson, B. Wynne, and J. Stilgoe, *The Public Value of Science: Or How to Ensure that Science Really Matters* (London: DEMOS, 2005).

47. Dryzek, *Politics of the Earth*, 8. Dryzek offers a checklist of elements to analyze any discourse: (1) the basic entities recognized or constructed in the discourse, (2) the assumptions about relationships inherent in the discourse, (3) the agents that make up the discourse and their motives, and (4) the key metaphors (i.e., political symbols) and other rhetorical devices involved in the discourse and in advancing it over competing discourses. Formal study of these reveals how any discourse is structured, how it operates, and how competing discourses might work together to find common ground. See also Swaffield, "Contextual Meanings in Policy Discourse," 199–224; C. MacDonald, "The Value of Discourse Analysis as a Methodological Tool for Understanding a Land Reform Program," *Policy Sciences* 36 (2003): 15. Analysis of claims and counterclaims provides insights into social process. See H. D. Lasswell, "The Language of Power," in H. D. Lasswell et al., eds., *Language of Politics* (Cambridge: MIT Press, 1966), 3–19; and Lasswell and McDougal, *Jurisprudence for a Free Society*.

48. D. A. Schön, *Beyond the Stable State* (New York: W. W. Norton, 1971); D. A. Schön, *The Reflective Practitioner: How Professionals Think in Action* (New York: Basic Books, 1983); D. A. Schön, *Educating the Reflective Practitioner* (San Francisco: Jossey-Bass, 1987); D. A. Schön and M. Rein, *Frame Reflection: Toward the Resolution of Intractable Policy Controversies* (New York: Basic Books, 1994); C. Argyris and D. A. Schön, *Theory in Practice* (San Francisco: Jossey-Bass, 1974).

49. Roe, "Varieties of Issue Incompleteness and Coordination."

50. C. Walters, *Adaptive Management of Renewable Resources* (New York: Macmillan, 1986); C. S. Holling, *Adaptive Environmental Assessment* (London: John Wiley and Sons, 1978); C. Walters and C. S. Holling, "Large-scale Management Experiments and Learning by Doing," *Ecology* 71(6) (1990): 2060–68.

51. A. J. Hansen, R. Rasker, B. Maxwell, J. J. Rotella, J. D. Johnson, A. W. Parmenter, U. Langner, W. B. Cohen, R. L. Lawrence, and M. P. V. Kraska, "Ecological Causes and Consequences of Demographic Change in the New West," *BioScience* 52, no. 2 (2002): 151–62.

52. G. Rochlin, "Defining 'High Reliability' Organizations in Practice: A Taxonomic Prologue," in K. Roberts, eds., *New Challenges to Understanding Organizations* (New York: Macmillan, 1993), 11–32; G. Rochlin, "Reliable Organizations: Present Research and Future Directions," *Journal of Contingencies and Crises Management* 4, no. 2 (1996): 55–59; C. Demchak, "Tailored Precision Armies in Fully Networked Battlespace: High Reliability Organizational Dilemma in the Information Age," *Journal of Contingencies and Crises Management* 4, no. 2 (1996): 2; T. LaPorte, "High Reliability Organizations: Unlikely, Demanding, and at Risk," *Journal of Contingencies and Crisis Management* 4, no. 2 (1996): 60–71.

53. M. S. McDougal, H. D. Lasswell, and W. M. Reisman, "The World Constitutive Process of Authoritative Decision," in M. S. McDougal and W. M. Reisman, eds., *International Law Essays: A Supplement of International Law in Contemporary Perspective* (Mineola, N.Y.: Foundation Press, 1981), 191–286.

54. D. N. Cherney et al., "Understanding Patterns of Human Interactions and Decision-Making: An Initial Map of Podocarpus National Park, Ecuador," *Journal*

of Sustainable Forestry (in press). This section is based on this paper and references on arenas cited in Cherney et al.

55. McDougal et al., "World Constitutive Process of Authoritative Decision," 191–286.

56. Brunner et al., eds., *Finding Common Ground;* M. I. Honig, "Where's the 'Up' in Bottom-Up Reform?" *Educational Policy* 18, no. 4 (2004): 527–36; C. Weible, P. A. Sabatier, and M. Lubell, "A Comparison of a Collaborative and Top-Down Approach to the Use of Science in Policy: Establishing Marine Protected Areas in California, *Policy Studies Journal* 32, no. 2 (2004): 180–207.

57. McDougal et al., "World Constitutive Process of Authoritative Decision," 191–286.

58. Ibid.

59. G. Bateson, *Steps to an Ecology of Mind* (New York: Ballantine Books, 1972), 155.

60. Numerous social scientists have written on the importance of context and why it must be mapped and understood to improve natural resource management policy. The "principle of contextuality" and a guide to mapping context are examined by Clark, *Policy Process*, 29. The dimensions of context are often neglected or overlooked in management cases. See also G. Honadle, *How Context Matters: Linking Environmental Policy to People and Place* (Bloomfield, Conn.: Kumarian Press, 1999). The practical and theoretical basis of context has been well described in the literature. Examples of researchers and managers mapping context are in T. W. Clark, A. R. Willard, and C. M. Cromley, eds., *Foundations of Natural Resources Policy and Management* (New Haven: Yale University Press, 2000). The groundwork for this kind of mapping was developed in H.D. Lasswell, *A Pre-view of the Policy Sciences* (New York: American Elsevier, 1971), and Lasswell and McDougal, *Jurisprudence for a Free Society*.

61. Contextual mapping requires asking and answering the following questions. Seven variables are recognized as the basis for understanding any context.

Participants: Which individuals and organizations are participating in the policy issue in question? Who wants to participate? Who should participate?

Perspectives: What demands are participants making? What expectations do they have? On whose behalf are demands made, i.e., with what groups or beliefs do people identify themselves?

Situations: What is the "ecology" of the situation, that is, the geographic features of the problem and the spatial and temporal scales involved, for instance? Are there any crises? Which institutions are or should be involved? Is the situation organized or not, and is it appropriately organized?

Values: What "assets" do participants have in terms of power, wealth, skill, knowledge (enlightenment), affection, well-being, respect, and rectitude? Which are most important to which participants?

Strategies: How are these value assets being used? Are people's strategies educational, diplomatic, economic, or militant? Are these used persuasively or coercively? How are the eight values employed in each strategy being used?

Outcomes: What are the results of each interaction or overall decision activity? Who benefits and who is harmed in terms of which values or assets?

Effects: What institutions and practices are promoted and which set back? Are new practices institutionalized? Are other practices abandoned?

62. E. Castle, *Information: The Human Resource* (Washington, D.C.: Resources for the Future, 1981), 4.

63. G. D. Brewer and P. deLeon, *The Foundations of Policy Analysis* (Homewood, Ill.: Dorsey Press, 1983).

64. This section based on M. H. Arsanjani, *International Regulation of Internal Resources: A Study of Law and Policy* (Charlottesville: University of Virginia, 1981); M. S. McDougal, H. D. Lasswell, and I. A. Vlasic, *Law and Public Order in Space* (New Haven: Yale University Press, 1963); D. M. Johnston, *The International Law of Fisheries* (New Haven: Yale University Press, 1965); and M. H. Arsanjani, *International Regulation of Internal Resources* (Charlottesville: University Press of Virginia, 1981).

65. See S. A. Primm and T. W. Clark, "The Greater Yellowstone Policy Debate: What Is the Policy Problem?" *Policy Sciences* 29 (1996): 137–66. These authors examine arguments of proponents and opponents in the debate. They recommend integrated approaches to resolve differences, including prototyping on the ground.

66. See C. M. Cromley, "The Killing of Grizzly Bear 209: Identifying Norms for Grizzly Management," in Clark et al., *Foundations of Natural Resources Policy and Management;* C. Cromley, "Bison Management in Greater Yellowstone," in R. D. Brunner et al., *Finding Common Ground,* 126–58.

67. See J. A. Pritchard, *Preserving Yellowstone's Natural Conditions: Science and the Perception of Nature* (Lincoln: University of Nebraska Press, 1999); and A. L. Haines, *The Yellowstone Story: A History of Our First National Park* (Yellowstone National Park, Wyo.: Yellowstone Library Association and Colorado Associated University Press, 1977).

68. J. Schechter, *Sustaining Jackson Hole: A Community Exploration* (Jackson, Wyo.: Charture Institute, 2004); R. Rasker, "Rural Development, Conservation, and Public Policy in the Greater Yellowstone Ecosystem," *Society and Natural Resources* 6 (1993): 109–26; and R. Rasker and A. Hansen, "Natural Amenities and Population Growth in the Greater Yellowstone Region," *Human Ecology Review* 7, no. 2 (2000): 30–40. See Clark et al., *Coexisting with Large Carnivores.*

69. M. S. McDougal, "Legal Basis for Securing the Integrity of the Earth-Space Environment," *Journal of Natural Resources and Environmental Law* 8, no. 2 (1992–93): 177–207.

70. J. Lash, *Critical Environmental Issues for 2005* (Washington, D.C.: World Resources Institute, 2004); and J. G. Speth, *Red Sky at Morning: America and the Crisis of the Global Environment* (New Haven: Yale University Press, 2004).

71. T. L. Anderson and F. S. McChesney, eds., *Property Rights: Cooperation, Conflict, and Law* (Princeton: Princeton University Press, 2003); and B. Yandle, ed., *The Market Meets the Environment: Economic Analysis of Environmental Policy* (Lanham, Md.: Rowman and Littlefield, 1999).

72. M. T. Klare, *Resource Wars: The New Landscape of Global Conflict* (New York: Metropolitan Books, 2001); and M. T. Klare, *Blood and Oil: The Dangers and Consequences of America's Growing Petroleum Dependency* (New York: Metropolitan Books, 2004).

73. B. Lomborg, *The Skeptical Environmentalist* (Cambridge: Cambridge University Press, 2001); B. Lomborg, eds., *Global Crises, Global Solutions* (Cambridge: Cambridge University Press, 2004). See J. Diamond, *Collapse: How Societies Choose to Fall or Succeed* (New York: Viking Press, 2005); R. Wright, *A Short History of Progress* (Toronto: House of Anansi Press, 2004).

74. Y. Dror, *The Capacity to Govern* (London: Frank Cass, 2001).

75. B. S. Low, "Human Behavior and Conservation," *Endangered Species Update* 21, no. 1 (2004): 14–22; and B. S. Low, E. Ostrom, C. P. Simon, and J. Wilson, "Redundancy and Diversity: Do They Influence Optimal Management?" in J. Colding

and C. Folke, eds., *Navigating Nature's Dynamics* (Cambridge: Cambridge University Press, 2003), 83–114.

76. For example, take a single location: R. D. Brunner, A. H. Lynch, J. Pardikes, E. N. Cassano, L. Lestak, and J. Vogel, "An Arctic Disaster and Its Policy Implications," *Arctic* 57 (December 2004), 155–67; and A. H. Lynch, J. A. Curry, R. D. Brunner, and J. A. Maslanik, "Toward an Integrated Assessment of the Impacts of Extreme Wind Events on Barrow, Alaska," *Bulletin of the American Meteorological Society* 85 (February 2004): 209–21.

77. J. Schneider, *World Public Order of the Environment: Toward an International Ecological Law and Organization* (Toronto: University of Toronto Press, 1979).

78. J. Schneider, *World Public Order of the Environment: Toward an International Ecological Law and Organization*, 4.

79. R. B. Keiter, *Keeping Faith with Nature: Ecosystems, Democracy, and America's Public Lands* (New Haven: Yale University Press, 2003), 14.

80. Ibid., 14.

Chapter 6. Improving Leadership

1. Y. Dror, *The Capacity to Govern* (London: Frank Cass, 2001).

2. M. Douglas, *How Institutions Think* (Syracuse: Syracuse University Press, 1986); R. Westrum and K. Samaha, *Complex Organizations: Growth, Struggle, and Change* (Englewood Cliffs, N.J.: Prentice-Hall, 1984); and G. Morgan, *Images of Organizations* (Beverly Hills, Calif.: Sage Publications, 1986).

3. D. P. Warwick, *A Theory of Public Bureaucracy: Politics, Personality, and Organization in the State Department* (Cambridge: Harvard University Press, 1975), provides an excellent model and description of how bureaucracies operate.

4. R. Westrum, *Sidewinder: Creative Missile Development at China Lake* (Annapolis: Naval Institute Press, 1999).

5. Y. Dror, *Ventures in Policy Sciences* (New York: American Elsevier, 1971), 216; also see Dror, *Public Policymaking Reexamined* (Scranton, Pa.: Chandler, 1968).

6. For an introduction to this skill set, see T. W. Clark, "Developing Policy-Oriented Curricula for Conservation Biology: Professional Education and Leadership in the Public Interest," *Conservation Biology* 15, no. 1 (2001): 31–39.

7. Dror, *Capacity to Govern.*

8. M. S. McDougal, H. D. Lasswell, and I. Vlisic, *Law and Public Order in Space* (New Haven: Yale University Press, 1963).

9. See T. W. Clark, "Policy-Oriented Professionalism: A Unique Standpoint," in *The Policy Process: A Practical Guide for Natural Resource Professionals* (New Haven: Yale University Press, 2002), 111–26. This section draws on Dror, *Capacity to Govern.*

10. Dror, *Capacity to Govern.*

11. Many authors have talked about the need to integrate, among them are L. H. Gunderson, C. S. Holling, and S. L. Light, "Barriers Broken and Bridges Built: A Synthesis," in *Barriers and Bridges to the Renewal of Ecosystems and Institutions* (New York: Columbia University Press, 1995), 489–532; National Research Council, "Integrating Knowledge and Action," in *Our Common Journey: A Transition Toward Sustainability* (Washington, D.C.: National Academy Press, 1999), 275–332; and C. Folke, F. Berkes, and J. Colding, "Linking Social and Ecological Systems: Management Practices and Social Mechanisms for Building Resilience," in *Linking Social and Ecological Systems: Management Practices and Social Mechanisms for Building Resilience* (New York: Cambridge University Press, 2000), 414–36.

12. Such a framework is described and illustrated in T. W. Clark, A. R. Willard, and C. M. Cromley, eds., *Foundations of Natural Resource Policy and Management* (New Haven: Yale University Press, 2000); and Clark, *Policy Process*.

13. See H. D. Lasswell, *A Pre–view of Policy Sciences* (New York: American Elsevier, 1971) for these standards.

14. Clark et al., *Foundations of Natural Resources Policy and Management*.

15. Much has been written on creativity, including: H. D. Lasswell, "Constraints on the Use of Knowledge in Decision Making," in *Information for Action: From Knowledge to Wisdom* (New York: Academic Press, Inc., 1975), 161–69; H. D. Lasswell, "The Social Setting of Creativity," in H. H. Anderson, ed., *Creativity and Its Cultivation* (New York: Harper and Brothers, 1959), 203–221; H. D. Lasswell, "The Continuing Revision of Conceptual and Operational Maps," in H. D. Lasswell, D. Lerner, and J. D. Montgomery, eds., *Values and Development: Appraising Asian Experience* (Cambridge: MIT Press, 1976), 261–84; and H. D. Lasswell, "Sharing the Experience of Permanent Reconstruction: A Policy Sciences Approach," in *Essays on Modernization of Underdeveloped Societies* (Bombay: Thacker, 1971), 536–46.

16. Dror, *Capacity to Govern*, 67–68.

17. T. W. Clark and A. M. Gillesberg, "Lessons from Wolf Restoration in Greater Yellowstone," in V. A. Sharpe, B. Norton, and S. Donnelley, eds., *Wolves and Human Communities: Biology, Politics, and Ethics* (Washington, D.C.: Island Press, 2001), 135–49; T. W. Clark, "Interdisciplinary Problem Solving in Endangered Species Conservation: The Yellowstone Grizzly Bear Case," in R. P. Reading and B. J. Miller, eds., *Endangered Animals: A Reference Guide to Conflicting Issues* (Westport, Conn.: Greenwood, 2001), 91–108; T. W. Clark, "Wildlife Resources: The Elk of Jackson Hole, Wyoming," in J. Burger, E. Ostrom, R. B. Norgaard, D. Policansky, and B. D. Goldstein, eds., *Protecting the Commons: A Framework for Resource Management in the Americas* (Washington, D.C.: Island Press, 2001), 91–108; and S. M. Wilson and T. W. Clark, "Resolving Human–Grizzly Bear Conflicts: An Integrated Approach in the Common Interest," in K. S. Hanna and D. S. Slocombe, eds., *Integrated Resource and Environmental Management: Concepts and Practice* (New York: Oxford University Press, 2006), 137–63.

18. These dynamic changes and demands are discussed in: T. W. Clark, "Interdisciplinary Problem Solving: Next Steps in the Greater Yellowstone Ecosystem," *Policy Sciences* 32 (1999): 393–414; and S. A. Primm and T. W. Clark, "The Greater Yellowstone Policy Debate: What Is the Policy Problem?" *Policy Sciences* 29 (1996): 137–66. These make problem definition difficult, see D. Dery, *Problem Definition in Policy Analysis* (Lawrence: University of Kansas Press, 1984).

19. A. I. Goldman, *Epistemology and Cognition* (Cambridge: Harvard University Press, 1995). See also M. Michalko, *Cracking Creativity: The Secrets of Creative Genius* (Berkeley, Calif.: Ten Speed Press, 2001) and J. L. O'Connor and I. McDermott, *The Art of Systems Thinking: Essential Skills for Creativity and Problem Solving* (San Francisco: Thorsons, 1997).

20. As M. S. McDougal noted, an appropriate balancing process should represent "a moving line of compromise varying with problems and contexts." He went on to say that, on the one hand, "all free peoples have a common interest in the establishment and maintenance of an inclusive competence adequate to secure common values and designed both to protect democratic access by peoples to participation in decisions," but that, on the other hand, people also have a "common interest in the establishment and maintenance of an exclusive competence adequate to protect particular peoples from arbitrary external interference and oppression and to promote the greatest possible freedom for initiative, experiment, and diversity in the effective adaptation of policies to local contexts." M. S. McDougal, H. D. Lasswell, and W. M.

Reisman, "The World Constitutive Process of Authoritative Decision," in M. S. McDougal and W. M. Reisman, eds., *International Law Essays: A Supplement to International Law in Contemporary Perspective* (Mineola, N.Y.: Foundation Press, 1981), 58–59.

21. McDougal et al., "World Constitutive Process of Authoritative Decision," 58–59.

22. M. H. Arsanjani, *Internal Regulation of Internal Resources: A Study of Law and Policy* (Charlottesville: University Press of Virginia, 1981).

23. An excellent example, where nongovernmental leadership has moved forward to help citizens clarify policy goals in terms of community development and environmental conservation, is the work of Jonathan Schechter of the Charture Institute and Jason Wilmot and Lydia Dixon of the Northern Rockies Conservation Cooperative, in the Jackson, Wyoming, area. See J. Schechter and J. Wilmot, "Sustaining Jackson Hole," *NRCC (Northern Rockies Conservation Cooperative) News* 17, no. 2 (2004): 3; and www.nrccooperative.org/SHJ/report.htm.

24. T. W. Clark and M. Ashton, "Interdisciplinary Rapid Appraisals: The Ecuadorian Condor Bioreserve Experience," *Journal of Sustainable Forestry* 18, no. ⅔ (2004): 1–30.

25. A. Miller, *Environmental Problem Solving: Psychosocial Barriers to Adaptive Change* (New York: Springer, 1999).

26. M. Weber, T. Parsons translation, cited in W. M. Reisman, 1981, "Private Armies in a Global War System: Prologue to Decision," in McDougal and Reisman, *International Law Essays*, 172.

27. An excellent example is government leadership by Parks Canada, which is partnering with universities (Simon Fraser University in British Columbia and Yale University in Connecticut) and a nongovernmental organization (Northern Rockies Conservation Cooperative in Wyoming) in seeking to reduce conflict significantly and improve grizzly bear management policy. This prototypical project is led by Mike Gibeau of Parks Canada. Innovative Q workshop and problem-solving workshops are being used.

28. This section draws on Dror, *Capacity to Govern*, 63–223.

29. Ibid., xv.

30. See G. D. Brewer, "Methods for Synthesis: Policy Exercises," in W. C. Clark and R. E. Munn, eds., *Sustainable development of the biosphere* (New York: Cambridge University Press, 1986), 455–75; P. deLeon, 1975, "Scenario Designs: An Overview," *Simulation and Games* 6, no. 1 (1975): 39–60; and H. A. DeWeed, "A Contextual Approach to Scenario Construction," *Simulation and Games* 5, no. 4 (1974): 403–14.

31. N. Tichy, *Managing Strategic Change: Technical, Political, and Cultural Dynamics* (New York: John Wiley and Sons, 1983); see also I. Smilie and J. Hailey, *Managing for Change: Leadership, Strategy, and Management* (New York: Oxford University Press, 2001).

32. G. Starling, "Levers for Implementation," in *Strategies for Policymaking* (Chicago: Dorsey Press, 1988), 498–561.

33. United Nations Development Program, *Capacity Development for Sustainable Human Development: Conceptual and Operational Signposts* (New York: UNDP, 1995), 14.

34. Definition comes from H. J. Aslin, N. A. Mazur, and A. L. Curtis, *Identifying Regional Skill and Training Needs for Integrated Natural Resource Management Planning* (Canberra, Australia: Bureau of Rural Sciences, 2002), 11.

35. For a description of hard and soft systems, see Aslin et al., *Identifying Regional Skill and Training Needs*.

36. G. H. Dession and H. D. Lasswell, "Order under Law: The Role of the Advisor-Draftsman in the Formation of Code or Constitution," *Yale Law Journal* 65 (1955): 174–95. For further description of the professional role and the advisor-draftsman, see R. Muth and J. M. Bolland, "Social Context: A Key to Effective Problem Solving," *Planning and Changing* 14, no. 4 (1983): 214–25; H. D. Lasswell, "Diversity: Synthesis of Methods, in *A Pre-view of Policy Sciences* (New York: American Elsevier, 1971), 58–75, as well as chapters entitled "Professional Services: The Ordinary Policy Process," 76–97, and "Professional Services: The Constitutive Policy Process," 98–111, and "Professional Training," 132–59.

37. D. P. Hanna, *Designing Organizations for High Performance* (Reading, Mass.: Addison-Wesley, 1988). C. E. Larson and F. M. J. LaFasto, *Team Work* (Newbury Park, Calif.: Sage Publications, 1989). R. Heller, *Achieving Excellence* (New York: DK Publishing, 1999). K. K. Smith and D. N. Berg, *Paradoxes of Group Life: Understanding Conflict, Paralysis and Movement in Group Dynamics* (San Francisco: New Lexington Press, 1987); J. R. Hackman, ed., *Groups That Work (and Those That Don't): Creating Conditions for Effective Teamwork* (San Francisco: Jossey–Bass, 1990)).

38. Hanna, *Designing Organizations for High Performance*, 13.

39. A highly useful organizational model is D. A. Nadler and M. L. Tushman, "A Model for Diagnosing Organizational Behavior," *Organizational Dynamics* 9, no. 2 (Autumn 1980): 35–51. See also T. W. Clark and R. Westrum, "High Performance Teams in Wildlife Conservation: A Species Reintroduction and Recovery Example," *Environmental Management* 13, no. 6 (1989): 663–70; and T. W. Clark and J. Cragun, "Organizational and Managerial Guidelines for Endangered Species Restoration and Recovery Teams," in M. L. Bowles and C. J. Whelan, eds., *Restoration of Endangered Species: Conceptual Issues, Planning and Implementation* (New York: Cambridge University Press, 1994), 9–33.

40. P. M. Burgess and L. L. Slonaker, *The Decision Seminar: A Strategy for Problem-Solving*, Merschon Center Briefing Paper No. 1 (Columbus: Merschon Center of the Ohio State University, 1978), 1.

41. H. D. Lasswell, "Technique of Decision Seminar," *Midwest Journal of Political Science* 4 (1960): 213–36; H. D. Lasswell, *The Future of Political Science* (New York: Atherton, 1963); H. D. Lasswell, *The Analysis of Political Behavior: An Empirical Approach* (Hamden, Conn.: Archon, 1966); H. D. Lasswell, "Decision Seminars: The Contextual Use of Audiovisual Means in Teaching, Research, and Consultation," in R. L. Merritt and S. Rokkan, eds., *Comparing Nations: The Use of Quantitative Data in Cross-National Research* (New Haven: Yale University Press, 1966), 499–524.

42. Lasswell, *Pre-view of Policy Sciences;* Brewer, "Methods for Synthesis," 455–75; A. R. Willard and C. H. Norchi, "The Decision Seminar as an Instrument of Power and Enlightenment," *Political Psychology* 14 (1993): 575–606. Also see J. J. Gargan and S. R. Brown, "What Is To Be Done? Anticipating the Future and Mobilizing Prudence," *Policy Sciences* 26 (1993): 347–59, and S. R. Brown, "The Composition of Microcosms," *Policy Sciences* 5, no. 1 (1974): 15–27.

43. L. L. Cunningham, "Applying Lasswell's Concepts in Field Situations: Diagnostic and Prescriptive Values," *Educational Administration Quarterly* 17 (1981): 21–43. Also see J. M. Bolland and R. Muth, "The Decision Seminar: A New Approach to Urban Problem Solving," *Knowledge* 6, no. 1 (1984): 75–88; R. Muth, "The Decision Seminar: A Problem-Solving Technique for School Administrators," *Planning and Change* 18, no. 1 (1987): 45–60. R. Muth and J. M. Bolland, "The Social Planetarium: Toward a Revitalized Civic Order," *Urban Interest* 3, no. 2 (1981): 13–25.

44. Many authors have focused on these and offer solutions, or at least partial solutions. For example, Dror, *Public Policymaking Reexamined*, noted that improvements could come from changes in management policy making, knowledge, personnel, structure and process, input and stipulated output, and in the environment. T. W. Clark and S. C. Minta, *Greater Yellowstone's Future: Prospects for Ecosystem Science, Management, and Policy* (Moose, Wyo.: Homestead Publishing, 1994), suggested improving people, science/management, agencies, and policy processes. W. Ascher and R. Healy, *Natural Resource Policymaking in Developing Countries: Environment, Economic Growth, and Income Distribution* (Durham: Duke University Press, 1990), 188, noted that comprehensive planning and realistic alternatives and an appropriately centralized and decentralized approach could help. They recommended thinking about the issue of size and complexity, taking a multidisciplinary involvement, doing better evaluation, and enabling easier termination. Also, they noted that regional authorities and local governments could help, as well as public participation and some market mechanisms. Other authors have also offered suggestions over the years. See, for example, M. R. Auer, "Agency Reform as Decision Process: The Reengineering of the Agency for International Development," *Policy Sciences* 31, no. 2 (1998): 81–105; J. Berry, G. D. Brewer, J. C. Gordon, and D. R. Patton, "Closing the Gap between Ecosystem Management and Ecosystem Research," *Policy Sciences* 31, no. 1 (1998): 35–54; G. D. Brewer, ed., "The Theory and Practice of Interdisciplinary Work," *Policy Sciences* 32, no. 4 (1999): 315–429; and E. L. Miles, "Personal Reflections on an Unfinished Journey through Global Environmental Problems of Long Timescale," *Policy Sciences* 31, no. 1 (1998): 1–33.

45. Elsewhere I have written about how to make improvements (e.g., T. W. Clark, E. D. Amato, D. G. Whittemore, and A. H. Harvey, "Policy and Programs for Ecosystem Management in the Greater Yellowstone Ecosystem: An Analysis," *Conservation Biology* 5 (1991): 412–22; Clark and Minta, *Greater Yellowstone's Future*; Clark, "Interdisciplinary Problem Solving"; T. W. Clark, "Species Recovery as Policy Process: Shifting the Perspective," in *Averting Extinction: Reconstructing Endangered Species Recovery* (New Haven: Yale University Press, 1997), 167–187; Primm and Clark, "Greater Yellowstone Policy Debate."

Chapter 7. Improving Management Policy

1. C. E. Lindblom, *Inquiry and Change* (New Haven: Yale University Press, 1990), 244.

2. Ibid., 3.

3. W. Asher and R. Healy, *Natural Resource Policymaking in Developing Countries: Environment, Economic Growth, and Income Distribution* (Durham: Duke University Press, 1990), 159–80; T. W. Clark, *Averting Extinction: Reconstructing Endangered Species Recovery* (New Haven: Yale University Press, 1997), 171.

4. B. Twight, *Organizational Values and Political Power: The Forest Service Versus the Olympic National Park* (University Park: Pennsylvania State University Press, 1983). The Forest Service, Fish and Wildlife Service, and National Park Service have been described by J. N. Clarke and D. McCool, *Staking Out the Terrain: Power Differentials among Natural Resource Management Agencies* (Albany: State University of New York Press, 1996); R. K. Merton, *The Sociology of Science* (Chicago: University of Chicago Press, 1973); H. Kaufman, *The Forest Ranger* (Baltimore: Johns Hopkins University Press, 1959); T. J. Tipple and J. D. Wellman, "Herbert Kaufman's Forest Ranger Thirty Years Later: From Simplicity and Homogeneity to Complexity and Diversity," *Public Administration Review* 51, no. 5

(1991): 421–28; S. Swaffield, "Contextual Meanings in Policy Discourse: A Case Study of Language Use Concerning Resource Policy in the New Zealand High Country," *Policy Sciences* 31, no. 3 (1998): 199–224.

5. J. Q. Wilson, *Bureaucracy: What Government Agencies Do and Why They Do It* (New York: BasicBooks, 1989).

6. Twight, *Organizational Values and Political Power.*

7. Merton, *Sociology of Science.*

8. Kaufman, *Forest Ranger.*

9. Tipple and Wellman, "Herbert Kaufman's Forest Ranger Thirty Years Later."

10. Ibid., 423.

11. R. D. Brunner, "Problems of Governance," in R. D. Brunner, C. Colburn, C. Cromley, R. Klein, and E. A. Olson, eds., *Finding Common Ground: Governance and Natural Resources in the American West* (New Haven: Yale University Press, 2002), 12.

12. H. D. Lasswell and M. S. McDougal, *Jurisprudence for a Free Society: Studies in Law, Science, and Policy* (New Haven: New Haven Press, 1992).

13. Brunner, "Problems of Governance," 12–14.

14. M. S. McDougal, H. D. Lasswell, and W. M. Reisman, "The World Constitutive Process of Authoritative Decision," in M. S. McDougal and W. M. Reisman, eds., *International Law Essays: A Supplement to International Law in Contemporary Perspective* (Mineola, N.Y.: Foundation Press, 1981), 191–286.

15. M. S. McDougal and H. D. Lasswell, "The Identification and Appraisal of Diverse Systems of Public Order," in M. S. McDougal and W. M. Reisman, eds., *International Law Essays: A Supplement to International Law in Contemporary Perspective* (Mineola, N.Y.: Foundation Press, 1981), 15–42.

16. For the benefits of participation, see E. C. Poncelet, "Personal Transformation in Multistakeholder Environmental Partnerships," *Policy Sciences* 34, nos. 3–4 (2001): 273–301; U. Wagle, "The Policy Science of Democracy: The Issues of Methodology and Citizen Participation," *Policy Science* 33, no. 2 (2000): 207–23; and G. Hampton, "Environmental Equality and Public Participation," *Policy Sciences* 32(2) (1999), 163–174. Also see R. Chambers, *Whose Reality Counts? Putting the First Last* (London: Intermediate Technology Publications, 1997), 103–5.

17. W. H. Ulfelder, S. V. Poats, J. B. Recharte, and B. L. Dugelby, "Participatory Conservation: Lessons of the PALOMAP Study in Ecuador's Cayamabe-Coca Ecological Reserve," America Verde Working Paper No. 1b (Arlington, Va.: Nature Conservancy, 1997). J. J. West and T. W. Clark, "Mapping Stakeholder Capacity in the La Amistad Biosphere Initiative," *Journal of Sustainable Forestry* 22, no. ½ (2006): 35–48.

18. G. E. Machlis and D. L. Tichnell, *The State of the World's Parks: An International Assessment for Resource Management, Policy, and Research* (Boulder: Westview Press, 1985).

19. M. P. Pimbert and J. N. Pretty, *Parks, People, and Professionals: Putting Participation into Social Development,* United Nations Research Institute for Social Development, Discussion Paper No. 57 (Geneva, Switz.: UNRISD, 1995), 29–30.

20. S. A. Primm, "A Pragmatic Approach to Grizzly Bear Conservation," *Conservation Biology* 10, no. 4 (1996): 1026–35.

21. T. P. Duane, "Community Participation in Ecosystem Management," *Ecology Law Quarterly* 24, no. 4 (1997): 771–97.

22. Y. Haila, "Environmental Problems, Ecological Scales and Social Deliberation," in P. Glaasberger, ed., *Co-operative Environmental Governance* (Dordrecht, Neth.: Kluwer, 1998), 181.

23. M. D. Frost, "Managing Diseases in the Greater Yellowstone Ecosystem: Ecological, Social, and Political Challenges" (master's thesis, Boston College, 2001), 116–24.

24. T. P. Duane, "Community Participation in Ecosystem Management," *Ecology Law Quarterly* 24, no. 4 (1997): 771–97; and J. Bohman, *Public Deliberation: Pluralism, Complexity, and Democracy* (Cambridge: MIT Press, 1996).

25. M. Wells and K. Brandon with L. Hannah, *People and Parks: Linking Protected Area Management with Local Communities* (Washington, D.C.: World Bank, U.S. Agency for International Development, and World Wildlife Fund, 1992); C. Moseley, "New Ideas, Old Institutions: Environment, Community, and State in the Pacific Northwest" (Ph.D. diss., Yale University, 1999); and J. K. Berry, "From Paradigm to Practice: Public Involvement Strategies for America's Forests" (Ph.D. diss., Yale University, 2000). See also R. Keiter, "Greater Yellowstone's Bison: Unraveling of an Early American Wildlife Conservation Achievement," *Journal of Wildlife Management* 61, no. 1 (1997): 1–11; J. S. Dryzek, *Discursive Democracy: Politics, Policy, and Political Science* (New York: Cambridge University Press, 1990); and S. A. Primm and T. W. Clark, "Making Sense of the Policy Process for Carnivore Conservation," *Conservation Biology* 10, no. 4 (1996): 1036–45; P. deLeon, *Democracy and the Policy Sciences* (Albany: State University of New York Press, 1997).

26. See, for example, H. D. Lasswell, "Constraints on the Use of Knowledge in Decision Making," in *Information for Action: From Knowledge to Wisdom* (New York: Academic Press, 1975), 161–69; H. D. Lasswell, "The Social Setting of Creativity," in H. H. Anderson, ed., *Creativity and Its Cultivation* (New York: Harper and Brothers, 1959), 203–21; and H. D. Lasswell, "The Continuing Revision of Conceptual and Operational Maps," in H. D. Lasswell, D. Lerner, and J. D. Montgomery, eds., *Values and Development: Appraising Asian experience* (Cambridge: MIT Press, 1976), 261–84.

27. For an introduction and review of learning see L. S. Etheredge, *Can Governments Learn?* (New York: Pergamon Press, 1985); L. S. Etheredge, "Government Learning: An Overview," in S. Long, ed., *Handbook of Political Behavior*, vol. 2 (New York: Plenum Press, 1981), 73–161; L. Etheredge and J. Short, "Thinking about Government Learning," *Journal of Management Studies* 20, no. 1 (1983): 41–58; C. Argyris and D. Schön, *Organizational Learning: A Theory of Action Perspective* (Reading, Mass.: Addison-Wesley, 1978); H. Wilensky, *Organizational Intelligence: Knowledge and Policy in Government and Industry* (New York: Basic Books, 1967); T. W. Clark, "Learning as a Strategy for Improving Endangered Species Conservation," *Endangered Species Update* 19, no. 4 (2002): 114–18; and T. W. Clark, "Organizational Learning: Institutionalizing Ignorance," in T. W. Clark, *Averting Extinction: Reconstructing Endangered Species Recovery* (New Haven: Yale University Press, 1997), 122–35.

28. H. D. Lasswell, *A Pre-view of Policy Sciences* (New York: American Elsevier, 1971), 190. For additional information about prototypes, see R. L. Knight and T. W. Clark, "Boundaries between Public and Private Lands: Defining Obstacles, Finding Solutions," in R. L. Knight and P. B. Landres, eds., *Stewardship across Boundaries* (Washington, D.C.: Island Press, 1998), 175–92; S. Primm and S. M. Wilson, "Reconnecting Grizzly Bear Populations: Prospects for Participatory Projects," *Ursus* 15, no. 1 (2004): 104–14; D. A. Glick and T. W. Clark, "Overcoming Boundaries: The Greater Yellowstone Ecosystem," in Knight and Landres, *Stewardship across Boundaries*, 237–56; C. Cromley, "Bison Management in Greater Yellowstone," in Brunner et al., *Finding Common Ground*, 126–58, and other cases in this book; J. L. Taylor, "Developing Environmental Management from a Case-Study Base," *Environmental Conservation* 9 (1983): 261; J. Romm, "Policy Education for Professional Resource

Managers," *Renewable Resources Journal* 2 (1984): 16; T. W. Clark and A. R. Willard, "Learning about Natural Resources Policy and Management," in T. W. Clark, A. R. Willard, and C. M. Cromley, eds., *Foundations of Natural Resources Management and Policy* (New Haven: Yale University Press, 2000), 25 (also in this book see Clark and Willard, "Analyzing Problems and Finding Solutions," 32–46); H. D. Lasswell, "Current Studies of the Decision Process: Automation Versus Creativity," *Western Political Quarterly* 8, no. 3 (1955): 381–99; H. D. Lasswell, "Strategies of Inquiry: The Rational Use of Observation," in D. Lerner, ed., *The Human Meaning of the Social Sciences* (New York: Meridian, 1959), 89–113; and H. D. Lasswell, "Policy Problems of a Data-Rich Civilization," in *Proceedings of the 1965 Congress, International Federation for Documentation, 31st Meeting and Congress, Washington, D.C., October 7–16* (London: Macmillan, 1965), 169–74; and C. M. Cromley, "The Killing of Grizzly Bear 209: Identifying Norms for Grizzly Management," in Clark et al., *Foundations of Natural Resources Policy and Management*, 173–220.

29. Knight and Clark, "Boundaries between Public and Private Lands."

30. Primm and Wilson, "Re-connecting Grizzly Bear Populations."

31. For example, see T. W. Clark, G. N. Backhouse, and R. P. Reading, "Prototyping in Endangered Species Recovery Programmes: The Eastern Barred Bandicoot Experience," in A. Bennett, G. Backhouse, and T. Clark, eds., *People and Nature Conservation: Perspectives on Private Land Use and Endangered Species Recovery*, Transactions of the Royal Zoological Society of New South Wales (Chipping Norton, NSW, Australia: Surrey Beatty and Sons), 50–62.

32. Good examples of case studies resource management are those in Brunner et al., *Finding Common Ground.*

33. J. L. Taylor, "Developing Environmental Management from a Case-Study Base," *Environmental Conservation* 9 (1983): 261.

34. J. Romm, "Policy Education for Professional Resource Managers," *Renewable Resources Journal* 2 (1984): 16.

35. T. W. Clark and A. R. Willard, "Learning about Natural Resources Policy and Management," in Clark et al., *Foundations of Natural Resources Management and Policy*, 25. Also in this book see Clark and Willard, "Analyzing Problems and Finding Solutions," 32–46.

36. H. D. Lasswell, "Current Studies of the Decision Process: Automation Versus Creativity," *Western Political Quarterly* 8, no. 3 (1955): 381–99; H. D. Lasswell, "Strategies of Inquiry: The Rational Use of Observation," in D. Lerner, ed., *The Human Meaning of the Social Sciences* (New York: Meridian, 1959), 89–113; and Lasswell, "Policy Problems of a Data-Rich Civilization."

37. Cromley, "The Killing of Grizzly Bear 209," and Cromley, "Bison Management in Greater Yellowstone." Also Glick and Clark, "Overcoming Boundaries."

38. S. Consolo Murphy and B. Kaeding, "Fishing Bridge: Twenty-five Years of Controversy Regarding Grizzly Bear Management in Yellowstone National Park," *Ursus* 10 (1998): 385–93.

39. M. K. Landy, "The New Politics of Environmental Policy," in M. K. Landy and M. A. Levin, eds., *The New Politics of Public Policy* (Baltimore: Johns Hopkins University Press, 1995), 211.

40. M. Shapiro, "Of Interests and Values: The New Politics and the New Political Science," in Landy and Levin, *New Politics of Public Policy*, 18.

41. Landy and Levin, *New Politics of Environmental Policy*, 224.

42. R. Westrum and K. Samaha, *Complex Organizations: Growth, Struggle, and Change* (Englewood Cliffs, N.J.: Prentice-Hall, 1984); R. Westrum, *Technologies and Society: The Shaping of People and Things* (Belmont, Calif.: Wadsworth, 1990); R. Westrum, "Management Strategies and Information Failure" (paper presented at

the NATO Advanced Workshop on Failure Analysis of Informational Systems, August 1986, Bad Winsheim, Germany).

43. A lot has been written on learning from diverse perspectives. Important among these is L. H. Gunderson, C. S. Holling, and S. L. Light, "Barriers Broken and Bridges Built: A Synthesis," in L. H. Gunderson, C. S. Holling, and S. L. Light, eds., *Barriers and Bridges to the Renewal of Ecosystems and Institutions* (New York: Columbia University Press, 1995), 489–532; and C. Folke, F. Berkes, and J. Colding, "Linking Social and Ecological Systems: Management Practices and Social Mechanisms for Building Resilience," in C. Folke and F. Berkes, eds., *Linking Social and Ecological Systems: Management Practices and Social Mechanisms for Building Resilience* (New York: Cambridge University Press, 2000), 414–36.

44. Etheredge, *Can Governments Learn?* ix.

45. Ibid.

46. See the section entitled "Proposal: A National Institution for the Training of Policy Leaders," pp. 81–98 in H. D. Lasswell, "On the Policy Sciences in 1943," *Policy Sciences* 36 (2003): 71–98. See also H. D. Lasswell, "Conflict and Leadership: The Process of Decision and the Nature of Authority," in A. V. S. deReuck and J. Knight, eds., *Ciba Foundation Symposium on Conflict in Society* (London: A. Churchill, 1966); R. D. Brunner and A. R. Willard, "Professional Insecurities: A Guide to Understanding and Career Management," *Policy Sciences* 36, no. 1 (2003): 3–36; and other papers in this volume.

47. Among these are J. McCroskey and S. D. Eininder, eds., *Universities and Communities: Remaking Professional and Interprofessional Education for the Next Century* (Westport, Conn.: Praeger, 1998); D. A. Schön, *Educating the Reflective Practitioner* (San Francisco: Jossey-Bass, 1987); D. A. Schön and M. Rein, 1994, *Frame Reflection: Toward the Resolution of Intractable Policy Controversies* (New York: Basic Books, 1994). The writings of Harold Lasswell are especially insightful, see H. D. Lasswell, "Current Studies of the Decision Process: Automation Versus Creativity," *Western Political Quarterly* 8, no. 3 (1955): 381–99; H. D. Lasswell, "The Social Setting of Creativity," in H. H. Anderson, eds., *Creativity and Its Cultivation* (New York: Harper, 1959), 203–21; Lasswell, "Strategies of Inquiry"; Lasswell, "Policy Problems of a Data-Rich Civilization"; H. D. Lasswell, "Sharing the Experience of Permanent Reconstruction: A Policy Sciences Approach, in *Essays on Modernization of Underdeveloped Societies* (Bombay: Thacker, 1971), 536–46; Lasswell, "Constraints on the Use of Knowledge in Decision Making"; and Lasswell, "Continuing Revision of Conceptual and Operational Maps."

48. Personal and professional security and insecurity is a key variable in leadership. See S. R. Brown, ed., "Special Issue: Professional Insecurities," *Policy Sciences* 36 (2003): 1–70, and other papers in this volume.

49. I have carried out about fifteen workshops over the last ten years that involve some of these features. Some were invaluable to participants, who have retained experience and contacts to the present.

50. Brunner and Steelman, *Adaptive Governance.*

51. Ibid.

52. A. Hohl and T. W. Clark, "Best Practices: The Concept, an Assessment, and Recommendations," unpublished.

53. Lasswell and McDougal, *Jurisprudence for a Free Society,* 210.

Chapter 8. Transitioning toward Sustainability

1. See D. A. Schön and M. Rein, *Frame Reflection: Toward the Resolution of Intractable Policy Controversies* (New York: Basic Books, 1994).

2. J. Dryzek, *Discursive Democracy: Politics, Policy, and Political Science* (New York: Cambridge University Press, 1990), 55.

3. M. Gladwell, *The Tipping Point* (Boston: Little, Brown, 2000).

4. S. Dovers, "A Policy Orientation as Integrative Strategy," in S. Dovers, D. I. Sern, and M. D. Young, eds., *New Dimensions in Ecological Economics* (Cheltenham, U.K.: Edward Elgar, 2003), 102–17, lists common themes discussed in this book. These are supported in his other works: S. Dovers, "A Framework for Scaling and Framing Policy Problems in Sustainability," *Ecological Economics* 12 (1995): 93–106; S. Dovers, "Sustainability: Demands on Policy," *Journal of Public Policy* 16 (1997): 303–18. See also P. Schwartz, *The Art of the Long View* (New York: Doubleday, 1991).

5. S. Dovers, "Policy Orientation as Integrative Strategy."

6. Ibid.

7. See T. W. Clark, A. H. Harvey, R. D. Dorn, D. L. Genter, and C. Groves, eds., *Rare, Sensitive, and Threatened Species of the Greater Yellowstone Ecosystem* (Jackson, Wyo.: Northern Rockies Conservation Cooperative, 1989).

8. S. Dovers, "Policy Orientation as Integrative Strategy."

9. Ibid.

10. Examples of this point are in T. W. Clark, M. B. Rutherford, and D. Casey, eds., *Coexisting with Large Carnivores: Lessons from Yellowstone* (Washington, D.C.: Island Press, 2005).

11. For a discussion of the growing constitutive crises, see the first and last chapters in R. D. Brunner, C. Colburn, C. M. Cromley, R. Klein, and E. A. Olson, eds., *Finding Common Ground: Governance and Natural Resources in the American West* (New Haven: Yale University Press, 2002).

12. L. J. Lundgren, ed., *Knowing and Doing: On Knowledge and Action in Environmental Protection* (Stockholm: Swedish Environmental Protection Agency, 2000); R. J. Sternberg, *Why Smart People Can Be So Stupid* (New Haven: Yale University Press, 2002); and S. W. Rosenberg, *The Not So Common Sense: Differences in How People Judge Social and Political Life* (New Haven: Yale University Press, 2002).

13. J. A. Merkle, "Scientific Management," in J. M. Shafritz, ed., *International Encyclopedia of Public Policy and Administration,* vol. 4 (Boulder: Westview Press, 1998), 2036–40.

14. M. K. McBeth and E. A. Shanan, "Public Opinion for Sale: The Role of Policy Markets in Greater Yellowstone Policy Conflict," *Policy Sciences* 37 (2004): 319–38.

15. S. Jasanoff, "NGOs and the Environment: From Knowledge to Action," *Third World Quarterly* 18, no. 3 (1997): 579–94.

16. L. Caldwell, cited in T. W. Clark, *Ecology of Jackson Hole, Wyoming* (Salt Lake City: Paragon Press, 1981), 97.

17. World Commission on Environment and Development, *Our Common Future* (Oxford: Oxford University Press, 1987).

18. See T. L. Friedman, *A Brief History of the Twenty-First Century* (New York: Farrar, Straus and Giroux, 2005).

19. L. W. Milbrath, "Sustainability," in R. Paehlke, ed., *Conservation and Environmentalism: An Encyclopedia* (New York: Garland, 1995), 612–13; L. W. Milbrath, *Envisioning a Sustainable Society: Learning Our Way Out* (Albany: State University of New York Press, 1989); L. W. Doob, *Sustainers and Sustainability: Attitudes, Attributes, and Actions for Survival* (Westport, Conn.: Praeger, 1995).

20. N. Mirovitskaya and W. Ascher, *Guide to Sustainable Development and Environmental Policy* (Durham: Duke University Press, 2001), 103.

21. Y. Dror, *The Capacity to Govern* (Portland, Ore.: Frank Cass, 2001), 39.

22. See J. A. Weiss, "The Powers of Problem Definition: The Case of Government Paperwork," *Policy Sciences* 22 (1989): 91–121, for an excellent description of how problem definition can shift policy process.

23. H. D. Lasswell, "The Emerging Conception of the Policy Sciences," *Policy Sciences* 1 (1970): 3–14.

24. R. D. Brunner and T. A. Steelman, *Adaptive Governance: Integrating Science, Policy, and Decision Making* (New York: Columbia University Press, 2005), 273.

25. M. Rolen, ed., *Challenges in Environmental Human Dimension Research* (Stockholm: Swedish Council For Planning and Coordination of Research, 1996), and M. Rolen, ed., *Culture, Perceptions, and Environmental Problems: Interscientific Communication on Environmental Issues* (Stockholm: Swedish Council For Planning and Coordination of Research, 1996).

26. M. Thompson, "Good Science for Public Policy," in M. Rolen, ed., *Culture, Perceptions, and Environmental Problems: Interscientific Communication on Environmental Issues* (Stockholm: Swedish Council For Planning and Coordination of Research, 1996), 10.

27. U. Svedin, " 'Human Dimensions' Research: Contemporary Challenges," in M. Rolen, ed., *Culture, Perceptions, and Environmental Problems: Interscientific Communication on Environmental Issues* (Stockholm: Swedish Council For Planning and Coordination of Research, 1996), 13.

28. Schön and Rein, *Frame Reflection,* 209.

29. Lasswell, "Emerging Conception of the Policy Sciences," 3–14; W. Parsons, *Public Policy: An Introduction to the Theory and Practice of Policy Analysis* (Cheltenham, U.K.: Edward Elgar, 1995); R. D. Brunner, "A Milestone in the Policy Sciences," *Policy Sciences* 29 (1996): 45–68; R. D. Brunner, "Introduction to the Policy Sciences," *Policy Sciences* 30 (1997): 191–215; R. D. Brunner, "Teaching the Policy Sciences: Reflections on a Graduate Seminar," *Policy Sciences* 30 (1007): 217–31; R. D. Brunner, "Raising Standards: A Prototyping Strategy for Undergraduate Education," *Policy Sciences* 30 (1997): 167–89.

30. H. D. Lasswell, "From Fragmentation to Configuration," *Policy Sciences* 2 (1971): 439–46.

31. L. Chen, *An Introduction to Contemporary International Law: A Policy-Oriented Perspective* (New Haven: Yale University Press, 1989), x.

32. This outline and descriptions rely on M. S. McDougal, H. D. Lasswell, and L. Chen, *Human Rights and World Public Order: The Basic Policies of an International Law of Human Dignity* (New Haven: Yale University Press, 1980), 399–415. Also see M. E. Caldwell, "The Community Power Process: An Inquiry for Policy-oriented Inquiry," in. H. D. Lasswell and M. S. McDougal, *Jurisprudence for a Free Society: Studies in Law, Science, and Policy* (New Haven: New Haven Press, 1992), 1439–88; and M. S. McDougal, "Human Rights and World Public Order: Principles of Context and Procedure for Clarifying General Community Policies," in Lasswell and McDougal, DH*Jurisprudence for a Free Society,* 1558–62.

33. V. C. Arnspiger, *Personality in Social Process: Values and Strategies of Individuals in a Free Society* (Chicago: Follett, 1961).

Appendix 1. Official Goals of the Greater Yellowstone Coordinating Committee

1. Greater Yellowstone Coordinating Committee, *GYCC Meeting Notes* (Bozeman: Greater Yellowstone Coordinating Committee, April 1994). This document

can be obtained from the executive coordinator, GYCC, c/o U.S. Forest Service, Bozeman, Mont., 59715.

2. Greater Yellowstone Coordinating Committee, *Greater Yellowstone Coordinating Committee Briefing Guide* (Billings, Mont: Greater Yellowstone Coordinating Committee 2001), 2.

3. Greater Yellowstone Coordinating Committee, *Summary Notes: Meeting with GYCC Assistant Secretaries Lyons and Frampton* (Jackson, Wyo.: Greater Yellowstone Coordinating Committee, September 24, 1994). This document can be obtained from the executive coordinator, GYCC, c/o U.S. Forest Service, Bozeman, Mont., 59715.

4. Greater Yellowstone Coordinating Committee, *Charter: Unit Managers' Team of the Greater Yellowstone Coordinating Committee* (Bozeman: Greater Yellowstone Coordinating Committee, April 1996). This document can be obtained from the executive coordinator, GYCC, c/o U.S. Forest Service, Bozeman, Mont., 59715.

5. May 1997 minutes, closed to public.

6. H. D. Lasswell, *A Pre-view of Policy Sciences* (New York: American Elsevier, 1971).

7. Greater Yellowstone Coordinating Committee, *Briefing Guide,* 2.

8. Ibid., 2.

9. U.S. Government, Memorandum of Understanding between the Rocky Mountain Region National Park Service, the Northern, Rocky Mountain and Intermountain Region Forest Service, and the Mountain Prairie Region Fish and Wildlife Service (Bozeman: Greater Yellowstone Coordinating Committee, 2003). This document can be obtained from the executive coordinator, GYCC, c/o U.S. Forest Service, Bozeman, Mont., 59715.

Index